"Books on leadership are many, but none are as
the Summit. Sure to be of value to every level of n.
valuable to leaders in government, in businesses of every size, and in every boardroom.
The uniquely experienced, gifted, and tested authors have led, inspired, and mentored
thousands in their extraordinary careers of service to our nation. They lived, observed, and
led from the most junior ranks to the pinnacle of military service—they walked the talk.
This book should be at the very top of every leader's list, to be read and reread."

— **Adm. Gary Roughead, U.S. Navy (retired)**
Chief of Naval Operations (2007–2011)

"Six outstanding American military leaders served our nation admirably and now share
their life experiences and the leadership lessons they learned. This book is a must-read for
all service members—both enlisted and officer—aspiring to be successful leaders in uniform
and beyond. Thank you to each of the authors for selflessly sharing their experiences and
insights on leadership and life."

— **Gen. Frank J. Grass, U.S. Army (retired)**
27th Chief of the National Guard Bureau (2012–2016)

"These extraordinary enlisted leaders have 'walked the talk,' and we should listen. Senior
noncommissioned officers are the backbone of the finest armed forces in the world. This
book shows why."

— **Adm. Thad W. Allen, U.S. Coast Guard (retired)**
23rd Commandant of the U.S. Coast Guard (2006–2010)

"A collection of personal vignettes that teach practical leadership lessons, *Breaching the
Summit* is a must-read for leaders at all levels. The authors have served as their services'
highest-ranking noncommissioned officers and are the epitome of the U.S. military's
professional, all-volunteer force."

— **Gen. Ed Eberhart, U.S. Air Force (retired)**
Commander, NORAD/USNORTHCOM (2002–2005)
CEO, Armed Forces Benefit Association

"The stories in this book are riveting, powerful, and, best of all, true. The leaders who
share their stories shaped our armed forces to be the greatest in the history of our nation.
I had the privilege of working very closely with Jim and Paula Roy for over two years. They
represent the very finest examples of great parents, humble and inspirational leaders, and
compassionate partners, and they showed us all how to have fun while performing at the
very highest levels of command."

— **Adm. Timothy J. Keating, U.S. Navy (retired)**
Commander, NORAD/USNORTHCOM (2004–2007), USPACOM (2007–2009)

"This book is a catalyst to renew the reader's commitment to reach and exceed their
potential—personally and professionally—while also taking joy in fulfilling their respective
duties and responsibilities."

— **Col. Paul H. Atterbury, U.S. Marine Corps (retired)**
Judge Advocate (1994–2014)

"I found this book inspiring, interesting, and instructive. Six remarkably accomplished authors share their powerful personal narratives, and their diverse and compelling stories lift up essential leadership lessons for us all, whether military or civilian. If you are anywhere along the continuum of leadership—from leading yourself to leading organizations—and want to become a better leader, read this book!"

— Vice Adm. Manson K. Brown, U.S. Coast Guard (retired)
   Assistant Secretary of Commerce (2010–2014)

"Of all the books I've read on military leadership, this is one of the absolute best. It's a great opportunity to learn from some of our nation's top enlisted leaders—a must-read for all service members."

— Chief Master Sgt. W. Allen Usry, U.S. Air Force (retired)
   Command Senior Enlisted Leader, NORAD/USNORTHCOM (2009–2011)

"*Breaching the Summit* is a motivating and inspiring read. The life lessons and stories shared will serve readers from any walk of life. The wisdom of these senior enlisted leaders in our U.S. military, tempered with their great sense of humility, reaffirms what makes our nation great. The advice these remarkable patriots have given has personally and professionally benefited me."

— Chief Master Sgt. James A. Cody, U.S. Air Force (retired)
   17th Chief Master Sergeant of the Air Force (2013–2017)

"*Breaching the Summit* is a must-read for those in leadership, those striving to become leaders, or those who want to become a better version of themselves. You will learn traits you want to possess, traits you already have and have forgotten, and traits that will guide you through your own leadership. You will connect and reminisce through the stories that helped mold these average individuals to become six of our top enlisted military leaders. Learn from their lessons of leadership as you become familiar with who they were before they became who they are now."

— Pam Swan
   Director, Military Relations for Veterans United Home Loans

"Brilliantly illustrating the concept 'grow where you're planted,' *Breaching the Summit* highlights six disparate individuals' adaptation of the basic tenets of leadership on their journey to the top of their respective services. Whether you are just beginning your own journey, or merely searching for additional inspiration to refine your own leadership style, the principles stressed in these pages provide a solid foundation for success."

— Force Master Chief Johnny Walker, U.S. Navy (retired)
   Naval Education and Training Command (2007–2009)
   MCPON Executive Assistant (2009–2011)

"Written by proven senior enlisted leaders, *Breaching the Summit* is a compendium of personal and professional experiences that distills success for all who choose the life of servant leadership in our military."

— Command Sgt. Maj. Mark Ripka, U.S. Army (retired)
   Command Senior Enlisted Leader, USJFCOM/USAFRICOM (2007–2011)

"*Breaching the Summit* is an insightful, informative walk through the phases of an enlisted career from some of the most successful leaders in modern history. This is a must-read primer for anyone on a journey through the ranks of our glorious military. Read this book. Learn from the real-life lessons that molded some of our greatest enlisted leaders."

> — Chief Master Sgt. John M. Harris, U.S. Air National Guard (retired)
> Command Chief Master Sergeant, Louisiana ANG (2006–2010)
> President, EANGUS

"I thoroughly enjoyed this book. This is not just for the 'seasoned' leader—anyone can relate to these lessons, no matter their level of leadership. I felt as though I were sitting in a Leadership 101 symposium!"

> — Fleet Master Chief April Beldo, U.S. Navy (retired)
> Manpower, Personnel, Training, and Education Command (2013–2017)

"As I'm passing 10 years since I left my own naval service behind, I am amazed at how the leadership truths in this book apply to any industry, not just the military. Making sure everyone knows their individual role in the unit's success, take the time to get the little things right, coach each other up, down, and all around, and all the other lessons. MCPON West said it best: don't just take care of your crew, challenge them to grow and be ready to take the reins. This book is not just for practitioners of the art of war—it is for anyone who wants to lead their organization to excellence!"

> — Capt. Bob Schuetz, U.S. Navy (retired)
> Deputy Commander and Chief of Staff, COMSUBPAC (2006–2008)
> Plant GM, Columbia Generating Station

"CMSAF Roy's anecdote of re-digging the water line in Michigan at age 10 was a foretaste of his sense of duty and doing the right things right throughout his AF career. His is a life well lived that resulted in Airmen being well led—including this one."

> — Lt. Gen. Loren M. Reno, U.S. Air Force (retired)
> Deputy Chief of Staff, USAF Logistics, Installations and Mission Support (2009–2012)

"MCPOCG Skip Bowen has a truly remarkable legacy of principled leadership built over an incredible career in the Coast Guard. His story, and the stories of the other leaders who reached the top positions in our military, are inspiring examples of linking personal and organizational success from the entry level to the summit. This book is a great read!"

> — Vice Adm. David P. Pekoske, U.S. Coast Guard (retired)
> Vice Commandant, USCG (2009–2010)
> Administrator, TSA

"What an amazing opportunity to read about these great leaders! MCPON Rick West's story shines light onto his sacrifice and compassion for the U.S. Sailor. Mentor, motivator, and total compassion describe MCPON West. His understanding of tactical and strategic level of knowledge provided excellence across the fleet. Most importantly, he is a chief's chief who connected and continues to connect with the CPO Mess at large."

> — Fleet Master Chief Susan Whitman, U.S. Navy (retired)
> U.S. Pacific Fleet (2015–2018)

"This book offers exceptional insight into what has shaped some of the finest senior enlisted leaders in modern history. Through easy-to-follow stories, *Breaching the Summit* showcases the impact small life events have in developing leaders. This team of authors has captured service-specific leadership challenges that have significance in today's joint warfighting environment, and I would consider *Breaching the Summit* a must-read for officers and senior enlisted looking to lead the military into the future."

— **Fleet Master Chief Mark Rudes, U.S. Navy (retired)**
Senior Enlisted Leader, PACOM (2012–2016)

"*Breaching the Summit* is a colorful and compelling look at the formative years, experiences, and philosophies behind our nation's preeminent senior enlisted leaders. America's faith in and empowerment of its enlisted corps has long been recognized as our true strategic advantage when it comes to military success at sea, in the air, and on land. Preston, Barrett, West, Roy, Jelinski-Hall, and Bowen show how humble beginnings and solid, consistent principles fueled their journeys and struck a chord with a generation of troops inclined to question the motives and methods of leadership. A must-read for anyone seeking insight on how to get the most out of people, or for anyone who doubts the tenacity of our young men and women wearing the uniform today."

— **Command Master Chief Scott Fleming, U.S. Navy (retired)**
Joint Task Force Guantanamo
MCPON Executive Assistant (2011–2013)

"From their humble beginnings to the pinnacle of their careers as the most senior noncommissioned officers of the U.S. Armed Forces, nowhere else will you find a compilation of lessons learned like those included here. These six senior enlisted leaders provide insights from which service members of all ages and ranks can learn, and that can serve as guides to a successful military career. I am proud to have served alongside these fine NCOs."

— **Command Sgt. Maj. Richard J. Burch, U.S. Army (retired)**
9th Command Sergeant Major of the Army National Guard (2010–2012)

"With stories from the lives and experiences of six of our top enlisted military leaders, *Breaching the Summit* offers the gift of leadership honing. This book is not only a must-read for current and future military leaders, but for those in the civilian sector as well. These warriors allow you to experience their personal stories of highs and lows on their way to the top enlisted leadership billet of each military branch. Though different in the challenges they faced, they are the same in telling the reason for their successes.

Patton once said: 'Your character is defined by how high you bounce when you hit bottom.' Each chapter illustrates how they took challenges head on and bounced to the pinnacle position of their respective professions. They suffered the pain of discipline to not suffer the pain of regret. I have added this to my professional library of leadership and urge all leaders—present and future—to do the same."

— **1st Sgt. Mark Gordon, U.S. Marine Corps (retired)**
Subject matter expert for combat profiling, Camp Lejeune

"Whenever a new initiative was considered, I knew no one had their finger on the pulse of the fleet like the MCPON and the SGTMAJ of the USMC. The troops see in these senior enlisted leaders a role model, parental figure, and most importantly, their potential best selves. Their hard-earned insights have value for any aspiring leader in any walk of life. HOOYAH!"

— Juan Garcia
17th Assistant Secretary of the Navy, Manpower and Reserve Affairs (2009–2016)

"*Breaching the Summit* is an amazing blend of leadership lessons from six very successful military leaders who started in humble beginnings and advanced to the highest levels of the enlisted corps. Their ability to weave personal and professional insights on their journey to mastering their tradecraft and leadership skills provides a blueprint of success for others. The evolutionary journey of learning leadership, practicing leadership, and then executing leadership is highlighted through their experiences. For those just starting their leadership journey and those who are seasoned leaders, there is much to be learned in this book."

— Chief Master Sgt. Steve McDonald, U.S. Air Force (retired)
Command Chief, Air Combat Command (2014–2017)

"It has been my pleasure to be associated with CMSAF James Roy for the last 25 years, beginning on the island of Guam, where he led the CE 'dirt boyz' in support of that critical infrastructure mission. Chief Roy possessed a keen insight into personal leadership then, which served our squadron well through five typhoon encounters and continued to serve him well throughout his successful military career, culminating as the Chief Master Sergeant of the Air Force. I'm proud to have played a small part in such a superb leader's life."

— Col. D. H. "Scott" Showers, U.S. Air Force (retired)
Commander, 36th Civil Engineer Squadron, Andersen AFB, Guam (1994–1996)

"*Breaching the Summit* gives a rare glimpse into the lives of leaders who served in the highest enlisted positions in the U.S. military. A must-read for any leader who'd like to make a difference in their organization—there's more common ground than you might think! Problem solvers, influencers, getting pushed out of their comfort zones, and passing it along were a few chunks that jumped out at me. Bravo on this collaboration of some AWESOME human beings!"

— Chief Master Sgt. Marty Klukas, U.S. Air Force (retired)
Command Senior Enlisted Leader, USTRANSCOM (2011-2014)

"MCPON Rick West's leadership style is more than a pillar from which many grew their naval careers—his motivation to challenge yourself is the foundation by which I continue to live and approach complex situations. His ability to build a cohesive, winning team is second to none. HOOYAH COB!"

— Machinist's Mate 1st Class Rich Hawkins, U.S. Navy (retired)
Captain, Delta Air Lines

"The U.S. military has long been a preeminent organization for producing leaders of exceptional capability and character. The authors of this book are among the best the military has to offer. It comes as no surprise that their thoughts would be insightful and well worth the attention of both aspiring and experienced leaders from all backgrounds and areas of interest."

— Mike Watt
CEO, Scientific Research Corporation

BREACHING THE SUMMIT

Inspiring Leadership Lessons from Six Military Journeys to the Top

Ken Preston	**Mike Barrett**	**Rick West**	**James Roy**	**Denise Jelinski-Hall**	**Skip Bowen**
SMA	SMMC	MCPON	CMSAF	NGB SEA	MCPOCG
Army	Marine Corps	Navy	Air Force	National Guard	Coast Guard

Breaching the Summit:
Inspiring Leadership Lessons from Six Military Journeys to the Top

First Printing, 2018

Editor: Elayne Wells Harmer

Remember My Service Productions
474 Bearcat Drive
Salt Lake City, UT 84115
www.remembermyservice.com

www.BreachingTheSummit.com

Library of Congress Control Number: 2018937014

Kenneth Preston, Micheal Barrett, Rick West, James Roy, Denise Jelinski-Hall, and Charles "Skip" Bowen. *Breaching the Summit: Inspiring Leadership Lessons from Six Military Journeys to the Top*. Salt Lake City: Remember My Service Productions, 2018.

ISBN 978-0-9863285-9-6

Printed in the United States of America

To book speaking engagements, send your request to *summitsixtalk@gmail.com*.

Dedicated to those who have so selflessly donned the cloth of this nation—past, present, and future—and to the devoted families who made their service possible.

Table of Contents

Table of Contents

Foreword

My first "senior" enlisted advisor was an E-5 Marine sergeant. I was a 22-year-old second lieutenant platoon leader in Vietnam in July 1968 when I lost my first Marine in combat to sniper fire from a nearby village. Lance Cpl. Guido Farinaro, a 19-year-old Italian immigrant from Bethpage, New York, was my machine gun squad leader. Guido attended a high school where everyone went on to college, but he chose to repay his adopted country and joined the Marines. I was holding Guido when he died. He was an incredible young Marine, and I was enraged. I called in an artillery strike on the village from which the sniper had fired. During the time between the call for fire and when the guns were set to fire the mission, my platoon sergeant, Sgt. Reid B. Zachary, now retired Sergeant Major Zachary, didn't say a word. He just looked at me in a way that let me know what I was doing was wrong. I called off the artillery strike, and we did what we should have done in the first place, which was sweep through the village on foot, where we found nothing but women and children. I honestly don't know how I could have lived with myself had I conducted that strike as my emotions almost had me do.

I've thought about that every day since. It taught me to set a moral compass—to think through who I am and who I want to be at the end of the day. It has also taught me the value of having a senior enlisted advisor—a battle buddy—to help guide and mentor certainly the young enlisted members of any unit, but also the officers and staff noncommissioned officers in positions of leadership.

To those outside the military, and even to those serving, the rank structure can sometimes be over simplified. It appears that we rack and stack everyone in the organization, and the person with the most rank "wins"—he or she is in charge, and everyone else has to follow orders that flow from the top. While there is certainly benefit in adhering to a chain of command, the interaction between the various ranks up and down that chain, officer and enlisted, becomes the connective tissue that creates a cohesive, successful organization inside of which good decisions and high morale thrive. The most senior member of a unit has the responsibility—and the honor—of leading. But to be successful, the planning and decision-making at the top must reflect a thorough understanding of the strengths at every level of the organization, for it is the enlisted leaders who will ultimately execute those plans in battle, and win.

Whatever success I achieved throughout my career has been in no small part to the senior enlisted leaders at every duty station, taking me under their wing, advising me in ways that made me a better leader, and more importantly, making the organization better. That is why I couldn't imagine taking on the responsibilities of Chairman of the Joint Chiefs of Staff without a senior enlisted advisor at my side. It was my great honor to have

Army Command Sergeant Major Joe Gainey as the first-ever Senior Enlisted Advisor to the Chairman. He was my eyes and ears, my sounding board, my battle buddy, and a solid anchor to hold to during the most challenging days as Chairman.

This book is written by six senior enlisted advisors who served similar roles for their respective service chiefs. Each shares his or her personal journey of hard work, sacrifice, and what can only be described as incredible success, but more importantly, the opportunity to lead others to success. Their "highlight reels" provide great examples to follow, but they have the humility to include the not-so-good decisions they made, and important lessons they learned along the way. We hear what drew them to the military, and after a lifetime of service in uniform, what struggles they faced as they transitioned following retirement. We have the opportunity to learn their lessons vicariously and apply them to our own circumstances, regardless of rank or position, in or out of the military.

This book is for junior service members new to the military who want a career path to emulate and solid advice about being an authentic leader, which requires more than just telling someone to get a haircut.

It is for senior enlisted leaders who want to compare notes, especially on how to best lead up, to help young officers understand what is expected of them, and to help senior leaders think through the challenges the organization will face.

It is a book for officers, to help them gain a better perspective on the vital role of the senior enlisted leaders in the organization. Early on in our careers, officers often focus on "who's senior." Over time, we learn to LEARN from everyone in our orbit, not just those who outrank us. This book demonstrates, early and often, this opportunity for growth and development learned from our senior enlisted.

This book is for families of our military as well. As service members, we often try to keep our warrior life separate from our home life; we don't want to over-share or burden our devoted family members with details that will cause additional concern. When we are home, we want to focus on family, and rightly so. But sometimes in our efforts to protect, we ask for buy-in and support without giving our families the opportunity to fully appreciate just what it is we do for a living. This book raises that curtain a bit, and lets our families see the other side. It also provides strong examples of the sacrifice, teamwork, and plain hard work required in any successful military family.

And finally, this book is for those who have no connection to the military whatsoever. Regardless of walk of life or current position—bottom of the ladder, top of the heap, or somewhere in mid-management—everyone can benefit from some candid advice from a battle buddy, such as those who wrote this book! ∎

— **Gen. Peter Pace, U.S. Marine Corps (retired)**
   16th Chairman of the Joint Chiefs of Staff (2005-2007)

Editor's Note

The English language—particularly the grammar rules governing spelling and punctuation—can seem complicated and nuanced. An editor's job is to ensure adherence to these rules while still making a book as reader-friendly as possible, all while staying true to the author's voice. Balancing these priorities is often tricky. In *Breaching the Summit*, I have done my best to merge military terms and traditional usage with English-language standards, but have had to make some trade-offs along the way that grammar purists will probably not appreciate. This is particularly evident in the formatting of ranks.

Members of the military are used to capitalizing a wide variety of terms, especially ranks and titles; to a military audience, seeing "sergeant major of the Army" in lowercase looks jarring and disrespectful. Such usage, however, is inconsistent with *The Chicago Manual of Style* (the primary style guide for fiction and non-fiction) as well as the style guides for each service. Nevertheless, I have attempted a compromise: ranks and titles are capitalized when they immediately precede a name (as is proper), or when it is the author's rank or title, or when it appears in the endorsements. In all other instances, ranks and titles are in lowercase.

Each service also has its own way of abbreviating ranks (e.g., "chief master sergeant" becomes "CMSgt" in the Air Force; "lieutenant commander" becomes "LCDR" in the Navy). For consistency, I have abbreviated ranks throughout the entire book in the traditional style per *Chicago*, but I have allowed service-specific acronyms (e.g., "PV2") as alternatives for variety. Per the style guide for military historians, I use an abbreviated rank when it precedes a full name ("Gen. George Washington") and a spelled-out rank on subsequent references when it precedes last name alone ("General Washington").

Whether you are a member of the military or a civilian, I hope you will see past any perceived inconsistencies in spelling and punctuation and focus on the authors' powerful messages. It's been an honor to help these exceptional Americans tell their stories. ∎

Kenneth O. Preston

SERGEANT MAJOR OF THE ARMY 2004–2011

K enneth O. Preston served as the 13th Sergeant Major of the Army from January 15, 2004, to March 1, 2011. He retired as the longest-serving SMA, with more than seven years in the position.

Preston is a native of Mount Savage, Maryland. He entered the Army on June 30, 1975, and attended Basic Training and Armor Advanced Individual Training at Fort Knox, Kentucky. Throughout his 36-year career, he served in every enlisted leadership position, from cavalry scout and tank commander to his final position as Sergeant Major of the Army. Other assignments he held as a command sergeant major were with the 3rd Battalion, 8th Cavalry Regiment, 1st Cavalry Division; 3rd "Grey Wolf" Brigade, 1st Cavalry Division; 1st Armored Division in Bad Kreuznach, Germany; and V Corps in Heidelberg, Germany. Prior to serving as SMA, Preston was the command sergeant major for Combined Joint Task Force 7 in Baghdad, Iraq.

As Sergeant Major of the Army, Preston served as the Army chief of staff's personal advisor on all Soldier and family-related matters, particularly areas affecting Soldier training and quality of life. He devoted the majority of his time in this position to traveling throughout the Army as a force provider for leaders at all levels of responsibility, overseeing Soldier and unit training, manning and equipping challenges, and talking to Soldiers and their families to understand their needs, personal hardships, and challenges serving a nation at war.

Preston's military education includes the Basic Noncommissioned Officer's Course, Advanced Noncommissioned Officer's Course, First Sergeant's Course, M1/M1A1 Tank Master Gunner Course, Master Fitness Trainer Course, Battle Staff Noncommissioned Officer's Course, and the United States Army Sergeants Major Academy. Preston holds a master's degree in business administration (MBA) from Trident University International.

His awards and decorations include the Distinguished Service Medal, Legion of Merit with oak leaf cluster, Bronze Star Medal, Army Meritorious Service Medal with three oak leaf clusters, Joint Service Commendation Medal, Army Commendation Medal with three oak leaf clusters, Army Achievement Medal with two oak leaf clusters, Good Conduct Medal 11th award, National Defense ribbon with bronze star, Southwest Asia Service Medal, Kosovo Campaign Medal, Global War on Terrorism Expeditionary Medal, Global War on Terrorism Service Medal, NCOES Ribbon, Army Service Ribbon, Overseas Service Ribbon 5th award, NATO Medal, Kuwait Liberation Medal (Government of Kuwait), Joint Meritorious Unit Award with bronze star, Army Meritorious Unit Commendation, and Department of the Army Staff Badge.

Preston serves as the vice president of Noncommissioned Officer and Soldier Programs at the Association of the United States Army (AUSA). He sits on the board of directors for United Services Organization (USO), Homes for Our Troops, Institute for Veterans and Military Families (IVMF) at Syracuse University, Army Historical Foundation, and Armor and Cavalry Heritage Foundation. He volunteers as the chief financial officer for Saint George's Episcopal Church in Mount Savage, Maryland. He also serves as vice-chairman of the Defense Advisory Committee on Women in the Services (DACOWITS) and is a member of the Army and Air Force Exchange Retiree Council.

Preston and his wife, Karen, have three children and eight grandchildren. They make their home once again in Mount Savage, Maryland. ■

Introduction: Soldier for Life

Things work out best for those who make the best of how things work out.
~ John Wooden

I grew up in rural western Maryland, a farming community surrounded by the Allegheny Mountains. My earliest memory is learning to ride my trike in our farmhouse kitchen when I was about two years old. After I conquered the tricycle, I gave it up to my younger sister and moved on to a three-wheeled kid tractor. Pretty soon I was riding on an old Ford tractor with my father, planting our garden in the spring and harvesting corn, potatoes, beans, and tomatoes in the fall.

Before I started first grade in 1963, we moved from the family farm to the outskirts of Mount Savage, about a mile from the school. Kids from first through twelfth grades all attended together; one of them, a pretty girl named Karen, became my wife many years later.

My parents also grew up in western Maryland, and both were veterans; my mother served in the Air Force from 1952 to 1955, and my father served in the Army from 1955 to 1957. They married in 1955, and their wedding pictures show them both in their service uniforms. Four uncles also served in the military, and though none of them talked about it much, I knew they were proud to be veterans.

The house my parents bought in Mount Savage was their first major purchase. I was only six, but I vividly remember overhearing their discussions about the price—$4,000 was a lot of money in 1963. On top of that, the old house needed renovation and the yard needed major landscaping. My parents were both young and did most of the renovations themselves, including painting, redesigning the interior, adding a new roof, and landscaping in the backyard that took years. I was the oldest of four, and my siblings and I helped as much as we could.

Each evening after working on the house, my father would lock himself away in the bedroom to study drafting, design, and sheet-metal mechanics. His efforts paid off as he moved from a laborer-type job to a skilled apprentice, then a tradesman. My father would make scaled models of the heating and air-conditioning ducting from paper and give them to me to use as toys, allowing me to link them together like building blocks. These early memories of my father—seeing him constantly developing and learning—had a lasting impact on my life. When I became a leader in the Army, I was determined to achieve my educational goals and make my Soldiers better than they were before coming under my care.

While my father gave me an expectation of lifelong learning, my mother's greatest gift was her optimistic outlook on life. Mom always looked for the silver lining in the tragedies, accidents, and stupid mistakes that affected our family. When I thought the end of the world had come because I broke my wrist in gymnastics and could not compete, or a summer camp field trip was cancelled, or one of my newspaper customers had skipped out of town without paying, she would say, "It's okay. Things always work out for the best."

She was always right. While there were missed opportunities or losses along the way, those unexpected incidents opened doors of opportunity for something new. I would never have had these opportunities but for those tragedies, accidents, and stupid mistakes. Even to this day, when faced with adversity and stress, I know in my heart that things will work out for the best.

The people of Mount Savage also gave me a gift. Every piece of that community had a positive impact on my outlook on life; they allowed me to be a kid and make mistakes, and helped me grow in wisdom. The Episcopal church where my mother's family attended (since the 1800s) played a key role in my development. One of my great-great-grandfathers, a Soldier with Company K of the 1st Maryland Volunteer Cavalry in the Civil War, is buried in the church's cemetery, as are my great-grandparents, grandmother, and grandfather, a doughboy in WWI.[1] Next to them are buried my parents, a brother, a brother-in-law, aunts, uncles, and cousins. Seeing my family tree, those who went before me, gave me an early perspective and appreciation that we are not here on this earth forever.

The community and the church played a huge part in establishing my values: love of family and the sacrifices we make to help each other. Biblical stories of Moses, Noah, Sampson, David and Goliath, and the life of Jesus molded me in my youth. I was an acolyte as a young boy and into my early teens, and sang in the church choir under the direction of Carol Berry, my devoted music teacher.

All the parts of my community were remarkably linked and supportive, like a marriage or a puzzle. I've weaved my experiences and values gained as a kid growing up in Hometown USA into my teachings, outlook, and visions as a leader. These fundamentals have served as my foundation and kept me focused on "what right looks like" in all the organizations and institutions where I have served and led.

I was an introvert in school—the quiet, shy kid who sat in the back, hoping the teacher would never ask me a question. I played Little League baseball in elementary school and intramural sports in high school, and even though I was a skinny kid of average athletic ability, I wanted to play centerfield for the New York Yankees. Gymnastics and baseball were my favorite sports, and I'm grateful to my coaches, George Bashields and Elwood Lashley, for their positive influence on a young boy.

My siblings and I had a happy childhood typical of rural life in the 1960s: we built treehouses in the yard and forts in the woods, and dug tunnels into the hillside to prepare for a Russian invasion or a zombie apocalypse. I loved my five-speed spider bike— ideal for doing wheelies and showing off—and my dream was to race Richard Petty in a NASCAR event or perform motorcycle stunts with Evel Knievel.

We didn't have a lot of money, so I learned early to adapt and improvise with what was available. We hardly ever threw things away—I fixed my own bicycle tires, built a ladder from scraps of wood and salvaged nails, and disassembled, cleaned, and repaired a lawnmower I found in the trash so I could mow lawns and earn money. Looking at

---

[1] According to H. L. Mencken, this nickname for service members in WWI can be traced to Continental Army Soldiers who kept the piping on their uniforms white through the application of clay. When the troops got rained on, the clay on their uniforms turned into doughy blobs.

my childhood from an adult's perspective, I realize that my ability to adapt and my determination to find solutions to my little-world problems were key attributes that would benefit my entire life. I was self-sufficient and dreamed big.

In junior high, I got a job delivering the *Cumberland Times*. I rode my bike for miles each day, dropping 170 newspapers into front storm doors before dawn and at night. My first purchase with my earnings was a new, straight-off-the-showroom-floor 1973 Harley Davidson 125cc street and trail motorcycle. My playground suddenly expanded to include the mountain trails and strip mines around Mount Savage, where you could ride a dirt bike all day and never follow the same trail twice.

My self-confidence grew as I entered my junior and senior years of high school. Getting my license to drive a car and motorcycle felt like a significant step forward, especially because my parents trusted me and treated me like an adult. Even when I drove too fast or tried to show off for friends, my parents allowed me to grow and learn from these lapses in judgment. They knew that natural consequences would teach me lessons that parental punishment couldn't.

As I started my senior year, the reality of life after high school began to settle in. I didn't have the grades for an academic scholarship or the talent for an athletic scholarship. My father offered to pay my college tuition, but I had watched him work multiple jobs for many years to provide for our family. I was determined to not be a financial burden.

In February of my senior year, I visited the Army recruiting office in Cumberland, Maryland, where Staff Sgt. James Underwood talked to me about occupational opportunities in the Army. Thinking I might follow in my father's footsteps as a draftsman, I dreamed of becoming an architect, designing and building homes. After taking the vocational aptitude test, my recruiter happily announced that based on my scores, I could have any job in the Army I wanted. I chose an engineering career field that would give me the opportunity to learn surveying, drafting, and design.

The next month, Staff Sergeant Underwood drove me to the Military Entrance Personnel Station (MEPS) for a physical, final screening, and the oath of enlistment. During the personnel screening part of my entrance exam, a civilian recruiter mentioned the need for Soldiers in the armor career field—and a $2,000 bonus for anyone who chose it.

Karen and I were planning to be married in a few months, and two grand was a lot of money in 1975. I didn't need much convincing—I changed my career field selection to armor crewman on the spot. I never regretted it.

Even though my dad was an Army veteran, my parents were not pleased that I had signed up. The Vietnam War was not a popular topic around my house (or anywhere, for that matter), and they worried for my safety. I assured them that my intent was to fulfill my four-year enlistment obligation, qualify for the Vietnam-era G.I. Bill,[2] and attend college to become an architect. With those plans firmly planted in my mind, Karen and I married at the end of our senior year, and I entered the Army on June 30, 1975.

If you had asked me then if I wanted to be a sergeant in the Army, my immediate

---

[2] The letters "GI" are often considered initials for "government issue," but they originally referred to equipment made from galvanized iron, such as trashcans used by the logistics services of the U.S. Armed Forces. "G.I. Joe" came to mean "U.S. Army Soldier" in American popular culture. The Vietnam-era G.I. Bill was a law that provided a range of benefits for returning veterans, including tuition and living expenses for college.

response would have been a resounding *no*. I never envisioned myself leading others, and I certainly didn't have the courage to commit to that type of responsibility. However, I have to admit that the idea of driving a tank and shooting a cannon seemed cool.

My first-ever airplane ride was the flight to Louisville, Kentucky, for basic training. Except for two weeks at summer camp in the sixth grade, I'd never been away from home. When I arrived at Fort Knox with 30 other new recruits, drill sergeants didn't waste any time capturing our attention, and we moved quickly into billeting at the reception station. After we'd each received a standard Army haircut and a new uniform, our drill sergeants began teaching us how to be Soldiers.

A Company, 11th Training Battalion, 5th Training Brigade was my first unit of assignment. The eight weeks of basic training seemed like an eternity as I longed for my new bride and my family back home, but the training kept us plenty busy, and we were dead tired by day's end with limited time to think. I wrote letters home every day, and the high point of my day was hearing my name during mail call. We were finally allowed to call home during the last few weeks of basic training.

My drill sergeant was Sgt. 1st Class Robert Daily. An infantry Soldier, Daily was older than the drill sergeants in the other platoons. His years of service and success as a noncommissioned officer and leader were evident by the decorations, service stripes, and combat patch on his service dress uniform. Sergeant Daily wasn't the loud, in-your-face drill sergeant you see in the movies, although he certainly could be loud when he needed to be. He had a stern look when he was serious, was direct in his guidance when he wanted us to do something, and was an expert in the skills he taught us. In his appearance, knowledge, fitness, and stamina, he was the epitome of a professional Soldier, and we all worked hard to emulate him.

Each morning, Sergeant Daily greeted us wearing a freshly starched uniform. During our intense physical training, we wore our uniform, trousers, T-shirts, and boots. We started early in the morning and marched for miles to the rifle and grenade ranges to begin training; in the evening, we took equally long road marches back to our billets.

I remember watching the sweat beads on Sergeant Daily's forehead and the wet spots seeping through his uniform as we trained in the heat of summer and the high humidity of Kentucky. Always erect, never slowing down, and hard as nails, he marched us to and from the ranges and ran us in the midday sun for physical training. Many times, he changed his uniform during lunch or dinner, and he always looked impeccable. His image and leadership style has stayed with me all these years.

I graduated from basic training at the end of August. Before starting the next eight weeks of Advanced Individual Training (AIT), I received a coveted four-day pass. I headed home as quickly as I could.

D Company, 3rd Training Battalion, 1st Training Brigade was home for the next phase of training. My drill sergeant was Sgt. 1st Class Lonnie Underwood, and, like Sergeant Daily, was an older, seasoned NCO. A Vietnam veteran who served in combat as a tank commander and platoon sergeant, Sergeant Underwood continued our demanding physical and intellectual training, beginning early in the morning and continuing until we went to bed.

We learned how to service and maintain the M60A1 Patton series tank, and tested our beginner driving skills through a hilly, muddy area. We learned gunnery skills by firing the tank cannon and machine guns, and I got to know my personal weapons, the .45-cal pistol and .45-cal submachine gun, like the back of my hand.

Overall, my 16 weeks in initial entry training were hard but positive, primarily because of what I learned from my two drill sergeants. My first lesson in what a leader should be, know, and do as a first-line supervisor came from watching Sergeants Daily and Underwood. Their maturity, knowledge, professionalism, and experience from years of dealing with thousands of different personalities from throughout the country made them the experts they were. I carried those lessons in leadership with me my entire Army career, and continued to learn from the professional examples of officers and noncommissioned officers.

After I successfully completed initial entry training, I earned my first promotion: from PV1 to PV2. I now had "mosquito wings" to proudly wear on my uniform, along with a nice pay raise.

I arrived at the battalion headquarters at Fort Hood, Texas, ready for my first duty assignment with the 1st Cavalry Division. The battalion's adjutant, a captain, glanced at my starched uniform and shiny boots.

"Private," the captain said, looking pleased, "would you like to be the driver of the battalion commander's Jeep?"

"No, thank you, Sir," I responded respectfully. "I joined the Army to serve on tanks, and I'd like to stay on a tank."

Sgt. Dale Stark escorted me to the Headquarters Company in the battalion, where I was assigned as the loader on the battalion commander's tank. Stark and his wife took Karen and me under their wing. The warm welcome they gave us—two teenagers now with a newborn daughter, 1,600 miles from home—demonstrated sponsorship and onboarding into the organization before either formal program existed.

After only five months in the Army, I knew the decision I had made to be part of this institution was right for my family and me. I was also determined to work hard and do my best to meet my leaders' expectations. Unknowingly, I had become a Soldier for life. ■

Lesson One: An Ideal Command Climate

The burden of establishing communication with the Soldier
rests upon the noncommissioned officer.
~ Sergeant Major of the Army Silas L. Copeland

In 1975, the 1st Cavalry Division was transitioning from an air-cavalry division in Vietnam to the ground-maneuver division it is today. My unit, Headquarters and Headquarters Company of the 2nd Battalion, 8th Cavalry, was an Army experimental unit that tested different concepts of force structure, training devices, and equipment modernization. We were in the middle of the Cold War—the tension that had been building between Eastern and Western powers since World War II—and the threat of the Soviet Union's aggression loomed like a dark cloud in the distance. We performed all our testing with that danger in mind.

We spent weeks at a time on the thousands of acres of training areas and firing ranges, and remained at approximately 70 percent manning levels during my time in the Headquarters tank section. After the first year, I rose to the rank of specialist assigned as the gunner on the battalion commander's tank. My driver and I often went to the field as a two-man crew, serving as a replacement or searchlight tank. It was a busy time for us, but I enjoyed what I was doing.

As a young, fledgling leader in my first permanent duty assignment, I paid close attention to the command heads. Each of them led by example—not just in their professional careers as leaders and Soldiers, but also in their personal lives as husbands and fathers. They didn't just focus on the success of each unit in the division, they also worked hard to get us home on the weekends and as many weeknights as possible to spend time with our families.

After a 36-year career, I still think of those leaders and what they taught me. My first tank commander and squad leader, Sergeant Stark; my company leaders, Captain Evanko and First Sergeant Thee; battalion commanders, Lt. Col. Thomas Buck and Lt. Col. William Hamilton; and division commander, Maj. Gen. Julius Becton all set the tempo for their units. Each of these men was a motivated, dedicated professional who empowered subordinates to make decisions to lead, teach, and control. They created a command climate of high expectations and trusted those of us in the lower ranks to rise to the challenge.

In the 1970s, life in the Army was tough for a young married couple with a baby. Soldiers at the rank of specialist and below with less than two years in the service were not permitted to live in government quarters. Karen and I initially lived in a small apartment in downtown Killeen, Texas, but the $135 rent per month was too high. For about a year, we lived outside of town in a small two-bedroom trailer. After two promotions and a pay raise the following year, we moved to a larger two-bedroom trailer in downtown Killeen.

Had it not been for the positive influence of my leaders on each level, I never would have stayed in the Army after my initial enlistment of four years. On reflection, I can attribute my re-enlistment at that first crossroads in my military career mainly to three things these early leaders taught me, and I practiced and taught these principles throughout my 36 years in the Army.

1. Positive command climate

Leaders and managers who serve at all levels—from the first-line supervisor on up to the first general officer or senior commander—are responsible for retention of their best and brightest. This is true in both the military and the civilian workforce.

The first reason Soldiers stay in the Army is command climate, the atmosphere where they live and work every day. An ideal command climate is a positive environment where Soldiers look forward to coming to work every day and doing their assigned jobs, where leaders at all levels are empowered and trusted to make decisions for their stewardship and to take control of their assigned piece of the Army. In an ideal command climate, supervisors mentor junior and senior leaders to ensure their organization is a success, and they give everyone the opportunity to contribute to the mission and the team.

2. Job satisfaction

This term means more than just enjoying your specific occupational specialty or chosen career field. It means enjoying what you do every day, regardless of where you are based or deployed. For me, having a positive supervisor who empowered me to do my work and manage my people was an essential piece of job satisfaction.

I had leaders who taught me my trade, explained how to make tough work a little easier, and showed me how to have fun at work. But most importantly, I felt valued and appreciated for my contributions to the team. I loved what I did, and I deeply respected my mentors for their efforts to make me feel better about the responsibilities I had to fulfill.

3. Quality of life

The third reason for staying in the Army—or any organization—is quality of life. The basic necessities for a healthy daily life are proper diet, rest, and exercise, not only for a single individual but also for their families. Being able to afford and access nutritious meals each day, having at least eight hours of rest, and having an exercise regimen that leaves you energized and feeling strong are the foundational factors for quality of life in a demanding career. Having time off from work to engage in hobbies or activities that allow you to reset yourself for the busy workweek is also a key factor.

As a young sergeant, I wanted to provide a quality of life for my family that was as good or better than what I could have provided back on the farm, living in a mobile home while working part time and going to school full time. I was grateful I could provide for them. Even though I had a demanding job that required a lot of time away from home, my family was safe, happy, and eager for me to return so we could spend time together.

As I moved upward into positions of increased responsibility, these three distinctive points of reference helped me assess retention success and identify the deterrents to retaining my best Soldiers. These early lessons helped shape my view of Army leaders and charted my opinions for the advice I would later give my commanders in each of my units. I have no doubt the success my commanders and I had in meeting our retention goals were attributed to the principles I learned in my first assignment with the 1st Cavalry Division. ∎

Lesson Two: The Promotion Process

Try not to become a person of success, but rather try to become a person of value.
~ Albert Einstein

During my time at Headquarters and Headquarters Company, the 2nd Battalion was shorthanded across the board in all occupational specialties. I was a young, hardworking armor crewman and scout. I was physically fit and motivated. I showed up to work on time and worked long hours to make the organization a success. Basically, I was a good Soldier.

The natural result of this situation is that I enjoyed early advancement to positions of increased responsibility. I arrived in the unit a PV2 (pay grade E-2)[3] straight out of advanced individual training. After four months in the unit, I was promoted to private first class, and four months later, when I met the minimum requirements for promotion to specialist, I was promoted again.

A couple of leadership lessons I learned during this time influenced my entire Army career, including how I viewed the value of employees and human capital in the civilian workforce. I was fortunate to have NCOs who saw my potential, valued my contributions, and empowered me with great responsibility—far beyond what would normally be expected of someone with my low rank and little experience.

My supervisors set high expectations, made sure I knew and understood the requirements, and then trained me for success. I was a proverbial "chunk of clay" coming out of initial entry training when I arrived in my unit. I was not the most physically fit, I was not the most knowledgeable tank loader or driver, and I was certainly not an expert in my profession. Thankfully, my first year in the Army was a leadership lab filled with hands-on experiences; I learned something new every day.

As I moved up through the ranks and took on increased responsibility, especially as I took on the role of leading and supervising other Soldiers, I looked to my leaders as role models. Sergeant Stark was an outstanding example. He didn't stand over his Soldiers and bark orders at them; he gave his Soldiers guidance and direction, focusing on areas that needed extra attention or help to get done as quickly, efficiently, and effectively as possible. After providing guidance, he rolled up his sleeves and jumped in too. He was the epitome of a working supervisor.

Fort Hood was a large installation with plenty of daily housekeeping chores. We cut grass and picked up trash along the highway adjacent to the post, but the most hated details of all were guard duty and manning the welcome center desk. In the early morning hours of January and February, walking guard around the helicopters parked on the airfield was miserable. Dressed in a field jacket, pile cap, and Army-issued gloves, and carrying a night stick to beat off any would-be vandals, these two-hour shifts seemed like

---

[3] Private second class. E-2 is the second-lowest pay grade in the Army's ranking hierarchy.

a lifetime of suffering.

My first battalion commander rarely occupied the position of tank commander, as he far preferred his Jeep. The only exception he made was during the tank gunnery crew qualification periods, when he had no choice but to be on the tank. Consequently, as a specialist and the gunner on the battalion commander's tank, I spent most of my time serving as tank commander.

During Basic Armor and Reliability Testing, the three tanks in the Headquarters Tank Section were divided among the line companies to augment their platoons. Over the course of approximately nine months, I spent most of my time with Sgt. 1st Class John "Pop" Tart in 1st Platoon, A Company. While the organization's equipment overall was old, my driver and I were proud of our tank and worked hard to service and maintain the 53-ton monster.

My crew and I had been in the field about 45 days when Sergeant Tart came to my tank one evening after we had finished eating. He handed me a small scrap of paper with numbers scribbled in ink.

"Specialist Preston, you have a promotion board tomorrow morning at 0800 at those grid coordinates," he said.

"Roger, Sergeant," was all I could say. "I will be there. Thank you."

I didn't sleep well that night. After starting the vehicles for stand-to at 0500, I prepared my tank for the movement. The grid coordinates were about a 45-minute drive from where we had spent the night. I arrived at the location known as Antelope Mound, put my tank in a defensive position, and prepared a range card. I put my driver in the gunner's seat to provide security while I pulled my last clean uniform out of a duffle bag, changed clothes, and shaved. At 0745 I grabbed my notebook, straightened my field gear, put my protective mask in place, and tightened my chinstrap before making my way to the small round tent at the top of the hill.

My battalion command sergeant major, my first sergeant, and a second first sergeant from one of the line companies were drinking coffee when I stepped into the tent and reported.

"Stand at ease, Specialist Preston," said the command sergeant major.

Given my nervousness, that was difficult. My first sergeant introduced me and described my duties and responsibilities. He gave them a rundown on my two years in the Army, my duty position as gunner and tank commander for the battalion commander, and my one assigned crew member, PV2 Benny Hare.

The CSM spoke up.

"Specialist Preston, why did you select the location where your tank was parked?" he asked.

"I parked the tank in a hull-down position," I answered, "facing the direction of the enemy with my gunner providing security."

"And what were your first actions when you occupied that position?" he continued.

"I prepared a range card for our fighting position, Sergeant Major."

He nodded, smiled, and shook my hand firmly. The others did, too.

"You're doing a great job, Specialist," he said. "You're dismissed to return to the fight."

That promotion board experience stayed with me my entire career. Two basic questions from the battalion CSM and a nod of approval from two first sergeants, then 90 days later—after only two years in the Army—I was promoted to sergeant.

I learned that if I demonstrated competency and had the confidence of my supervisors, I could get promoted. Although I had already been doing everything a sergeant was expected to do, I now wore the rank on my collar and moved up a pay grade. I had demonstrated the technical and tactical skills necessary to be successful and had the confidence of the leadership above me. ■

Lesson Three: Growing Leaders

Courage is resistance to fear, mastery of fear—not absence of fear.
~ Mark Twain

I n the Army, how we grow leaders is fundamentally a three-step process. I have used these principles to demonstrate not only the growth of young leaders in the Army at the lieutenant and sergeant levels, but also to show the link to discipline within an organization. This lesson became one of my most valuable teaching tools as a senior CSM and sergeant major of the Army.

1. Establish a standard

The first step for growing a first-line leader is to establish standards and make sure everyone knows them. You can't do a good job if you don't know what a good job is. The Army has lots of standards, and we use the term broadly. You will hear platoon sergeants say, "Enforce the standard!" First sergeants say, "Train to the standard!" The battalion CSM might say, "That is not our standard," when he gets called out by the brigade CSM for the empty beer cans in the parking lot.

Standards take the form of regulations, policies, expectations, procedures, or minimum requirements. Take, for example, Army Regulation 670-1: Wear and Appearance of the Uniform. The Department of the Army established this regulation, which provides guidance and policy on the wear of duty and service dress uniforms, as well as hair and grooming guidance. These regulations constitute an established set of standards for more than one million Soldiers in the regular Army, Army National Guard, and Army Reserve.

Another Army regulation provides procedures for the safe operation of every vehicle and every piece of equipment used in more than 140 different career fields. Preventive maintenance checks and services on these vehicles and associated equipment is expected to be performed in accordance with the operator's manuals, following a step-by-step process.

Many other standards in the Army are established at the small-unit levels. Unit standard operating procedures (SOP) outline the placement of ammunition pouches and equipment on the outer tactical vest (body armor), and even where the name on a helmet band is placed. SOPs exist for the charge of quarters and duty officers in the execution of their duties and security checks after hours. A unit SOP designates load plans for tools and equipment stored on each vehicle. Even the mop closet in the barracks hallway has a standard of cleanliness and organization that must be met by 0900 every morning.

2. Enforce the standards

The second step for growing an outstanding leader is to put someone in charge of enforcing the standards. This responsibility begins with a sergeant and the two to three Soldiers they are responsible to train, lead, professionally develop, discipline, and keep informed. The responsibility placed on the sergeant to train and enforce standards that apply to them and their Soldiers is critical for maintaining competent, confident, and disciplined organizations.

3. Hold supervisors accountable

Step three in this process calls for senior leaders to hold sergeants accountable for the Soldiers under their supervision. That means when Private Preston looks like a goat's butt and is not doing what he should be doing, it is Private Preston's sergeant who is held accountable. I learned this lesson and practiced it every day with the Soldiers in my care. They were my responsibility and stewardship—not only during the day, but also after duty and on the weekends.

Only when I became a senior leader did I fully understand and appreciate how this simple three-step process affects the discipline of large units.

Twenty years after leaving Fort Hood as a sergeant, I returned to the 1st Cavalry Division, this time as the brigade command sergeant major for the 3rd "Greywolf" Brigade. The division headquarters and 1st Brigade were deployed to Bosnia for the first stateside deployment of a unit into the Balkans. I had been a battalion and brigade CSM in the unit for more than three years. With one-third of the division deployed and the remaining two-thirds at Fort Hood, I found myself dual-hatted as both the brigade and division rear CSM. For the first time in my Army career, I had the opportunity to walk into and observe other large units outside my sphere of control.

Late one evening, an infantry company was conducting training on one of the live-fire ranges when a Soldier was killed. I received a call from the division rear commander informing me of the tragic death and asking me to join him early the next morning to meet with the firing unit's leadership. I met the senior commander at the division headquarters and joined him for the trip to the range.

As the CSM in a heavy infantry brigade, I was very familiar with this range—I had been there many times with my Soldiers for training. As we arrived at the range and drove through it, the first thing I noticed was the ammo pad. The Soldier on guard was sitting on the edge of the elevated concrete pad, feet dangling, helmet off, and looking like he'd had a long night. The ammo pad is essentially a loading dock that allows combat vehicles to pull alongside and upload or download allocated ammo. That day, belts of ammo for the 25mm chain gun on the IFV[4] were lying loose and bundled in piles on the concrete—right where the firing crews downloaded it when they transitioned from the range back to the parking area.

Like all military installations, Fort Hood has a specific set of range operating procedures that define the safety and management rules for an ammo pad. These rules include having a vigilant Soldier on guard, storing ammo off the concrete surface on four inches or more of dunnage (i.e., a pallet), covering ammo with a tarp to protect it from the elements, having a nearby fire extinguisher and warning signs, no smoking within 50 feet, and so on. As I glanced at the ammo pad, I made a mental note of the guard, uncovered ammo lying on the concrete, no visible fire extinguisher, and no warning signs.

We continued in the van to the control tower for the range complex. As we got out, a young captain company commander, two lieutenant platoon leaders, and two platoon

---

4 M2 Bradley Infantry Fighting Vehicle (an armored personnel carrier)

sergeants approached us. As I looked at these five senior leaders, I immediately noticed how they were dressed and how they carried themselves.

The company commander wore a holster for his 9mm pistol tied with 550 cord[5] upside down to his load-bearing equipment. The other four were all wearing their equipment differently. I made a mental note that this unit must have been unaware of the tactical operating procedures for how to wear or configure their field uniform equipment. There seemed to be no standard; the decision for what field gear to wear and how to wear it was apparently left to the individual.

We proceeded down the range road to the site where the Soldier had been killed, and stopped where the infantry teams and squads in the back of the IFVs had dismounted to clear the wood-line and a small bunker as part of the firing exercise the night before. When the dismounted infantry had moved to the top of the ridge, the four IFVs maneuvered onto the high ground to provide over-watching and supporting fire. Simulating an enemy force that had withdrawn or was forced to abandon the ridgeline, the dismounted infantry cleared and occupied the enemy's complex trench system and prepared for a counterattack of enemy forces. It was during this stage of the exercise that one of the four-man teams placed itself too far forward on the abutment of the trench, and a Soldier was accidentally shot and killed.

As I walked the terrain and listened to the debriefing, several things caught my attention. Both infantry platoons on the range had the opportunity to conduct the exercise without ammunition before going live-fire for both day and night scenarios. Clearing a complex trench system requires a set of standard operating procedures that all Soldiers on the ground are supposed to know and understand. Visual and voice communications between the Soldiers on each of the teams and between the teams clearing their assigned trenches are essential when you do not have direct visibility with all the players. These known and practiced operating procedures support the advanced techniques needed to quickly, efficiently, and safely clear complex trench systems and buildings with multiple rooms and levels.

It didn't take long to discover that the firing platoon had not followed established operating procedures, and the non-firing platoon that provided the observing safety personnel did not follow range operations to stop any unsafe acts. There on a ridgeline of a training range, I had one of those *aha!* moments. For years, senior NCO leaders had taught me to enforce the standard, train the standard, teach the standard, and be the standard, but I finally, fully understood how essential standards are to discipline and safety.

When sergeants stop enforcing standards at the fundamental level, a domino effect occurs that spirals downward until something tragic occurs in the unit—a piece of equipment is damaged or destroyed, or a Soldier is injured or killed. *Why did sergeants in this company of 150 Soldiers stop enforcing standards?* I thought to myself, baffled. After a few days of analyzing the situation, I determined that it had started at the top: the commander of that rifle company had, by his lax standards, unwittingly circumvented the authority of every sergeant in his company.

Sergeants are responsible for inspecting their Soldiers and expecting them to be in the correct uniform each day—that's a critical piece of empowering a new leader to take

5 Parachute cord with a minimum breaking strength of 550 pounds

charge and make corrections. When the company commander wears field gear that doesn't conform to the established standard, the sergeants are no longer empowered to make those uniform corrections on their Soldiers.

The domino effect from this fundamental failure is that sergeants then stop ensuring that Soldiers are clearing their weapons by the standard; they stop following the operator's manual when doing preventive maintenance checks and services on their equipment; they stop checking on their Soldiers during after-duty hours; and they stop making on-the-spot corrections when they see something out of place, whether it's a safety violation or inappropriate language. Instead, sergeants who have lost their authority choose the easier path and simply ignore the problem.

During the first year of my tour in Iraq, I learned that enforcing standards in training prior to a deployment is critical. Almost all large units were fragmented across the battlefield while executing their assigned missions. Even platoon leaders routinely executed five or six missions simultaneously, with their sergeants or staff sergeants and assigned Soldiers also performing different tasks. These platoon leaders were more appropriately platoon "commanders" who relied on the expertise of their junior NCOs and Soldiers to accomplish the missions with little or no oversight. Those units that did not develop competent and confident leaders in training, prior to deployment, struggled in the combat zone. It was too late for these units to get their act together—the operational tempo of the day-to-day missions in combat allowed little time for fixing problems, and these units struggled throughout the duration of their deployment. These were the units that needed a lot of love from the most senior sergeants major to get them back on track.

Enforcing standards at the first-line leader level is the foundational element to building a disciplined organization in which NCOs run day-to-day operations efficiently and effectively, allowing officers to plan for and command future operations. ∎

Lesson Four: Learning from Failure

Never give in, never give in, never, never, never, never–in nothing, great or small, large or petty–never give in except to convictions of honor and good sense.
~ Winston Churchill

I n January 1978, I left Fort Hood for a new assignment at the Army base in Gelnhausen, Germany. We put our few household goods into storage, and Karen and our two children stayed with her parents until I found a place to live. We were young and a little nervous about the move, but after two years at Fort Hood, I was confident I was an expert in my career field, with little left to learn.

In reality, of course, I was just a young grasshopper (think David Carradine in the TV series "Kung Fu") with a great deal to learn. The next three years in Germany would provide many lessons in leadership that shaped my views and personality as a leader for the rest of my Army career and life.

My unit was deployed when I arrived, so I spent the first week in Gelnhausen getting in-processed, re-establishing my pay location and the address of my unit, getting fitted and issued all my field clothing, and being assigned a temporary room in the barracks until I could obtain government quarters or lease an apartment off post. The day my company returned, I immediately reported to my first sergeant for duty. He welcomed me to the unit and told me I would be the charge of quarters that night to watch over the Soldiers in the barracks. This first night in the unit gave me the opportunity to read all the policy letters and operating procedures. I wanted to learn every standard and expectation of the organization.

My first impression of the unit was not positive. The attitude and outlook of the leadership and Soldiers were completely opposite those at Fort Hood. B Company occupied the second floor of an old German Army barracks building, likely built in the 1930s, and the company offices were cluttered, dirty, and depressing. The barracks weren't any more heartening, as I occupied a small room with three other NCOs from the company. My roommates were Staff Sgt. Ricardo "Robby" Robinson, a tank commander in 1st Platoon; Sgt. Chris Breckling, a turret mechanic; and Cpl. Michael Boaer, a gunner on the 3rd Platoon, platoon sergeant's tank. Our room was the last one at the end of the hallway, directly across from the latrine and the shower rooms. The good part about that location was a short walk to the bathroom–the bad part was the smell and the noise from everyone using the bathroom at all hours of the night. I missed my family and was anxious to find a place for us to live so they could join me in Germany.

I was relieved when the housing office found a basement apartment in a small German village about nine miles from the base. I quickly passed my driver's test and bought a 1964 German Ford Taurus for $150. It had a small oil leak, but some friends helped me repair it over the weekend. I had a car and a place to live, and my family was on their way–I was set!

Meanwhile, I was assigned as the gunner on the 1st Platoon, platoon leader's tank.

After being a gunner on a battalion commander's tank, working for the lieutenant was easy. There were five tanks in the platoon, B11 to B15; the platoon leader's tank was B11, and the platoon sergeant's was B14. For the first few weeks, I spent long hours cleaning the tank and putting it back together after the unit's return from training. It was important to prove my technical and tactical proficiency.

I had been in Germany for about 60 days when Karen and the children arrived. About a week later, my first sergeant sent me to the Basic Noncommissioned Officer Course in Vilseck, a town in northeastern Bavaria. This was the first of three times in my Army career when I would leave my family shortly after their arrival at a duty station, leaving them to move and set up the household without me. My neighbor, a sergeant in one of the infantry battalions, promised to look after Karen and the children while I was gone for three months.

After graduating from the NCO course, I packed my Class A service uniform in my duffle bag, donned my field gear, and headed to the ranges in Grafenwöhr just outside the back gate of Vilseck. My unit had already deployed from home station for a month-long tank gunnery training. En route to the range, I learned that my platoon leader had been compassionately reassigned to Fort Sam Houston, Texas, for a family emergency. I was now the tank commander of B11.

The reality of my situation sunk in quickly. I was 20 years old, fresh from the Basic NCO Course with only my Fort Hood training behind me. I had never commanded a tank on a crew qualification course, had not been part of the home-station training and preparation for this gunnery period, and none of my crew had trained for this monumental task. Making my situation even more difficult was the short time frame—the unit was already on Tank Table VI,[6] which at that time consisted of a series of firing exercises for a stationary tank to engage both stationary and moving targets with the main gun and tank-mounted machine guns.

Yet despite those challenges and my inexperience, I was undaunted and naively confident. Tank crew qualification on Tank Table VIII was only three days away, but I was up to the challenge. I moved my driver into the gunner's seat and recruited Sergeant Breckling, our company's turret mechanic, to be my driver. I was ready to command my tank and safely engage targets.

My self-assurance soon took a hit. Realistically, there simply wasn't enough time to practice and train a crew to the level of competency needed to successfully qualify seven of the ten engagements on Tank Table VIII. We didn't pass.

For the first time in my Army career, I had failed. I failed my Soldiers who were depending on me to make them a success, and I failed my leaders who were depending on me to make the unit a success. While all the senior leaders in the company understood the extenuating circumstances behind our failure to qualify, I lived under a cloud of shame and embarrassment until I could prove I was and could be a good tank commander. I was determined to do everything in my power to succeed if I were ever given the chance to command a tank again.

While I blamed myself for the failure on the gunnery range, I also found myself

---

6 Tank Tables consist of 12 qualification exercises for tank gunnery proficiency.

harvesting a degree of bitterness and anger toward my platoon leadership. The platoon sergeant and other tank commanders could have done a lot to help set me up for success before I ever arrived on the gunnery range. The platoon had had six weeks of home-station gunnery training and preparation followed by two weeks of preliminary gunnery training at Grafenwöhr while I was taking the NCO course, yet nothing was done to help prepare my crew or tank before I arrived.

All experiences are learning opportunities if you look closely at the circumstances leading up to the event. Analyzing all the contributing factors is the key to improvement. I hid my frustrations and kicked rocks in the parking lot while I considered what I could have done better. I realized I still had a lot to learn. ∎

Lesson Five: 11 Principles of Leadership

The function of leadership is to produce more leaders, not more followers.
~ Ralph Nader

A short time after that disappointing experience in Grafenwöhr, we received a new company commander and first sergeant: Capt. James E. Bessler and 1st Sgt. Gary P. Pastine. New leadership generally means changes in how an organization is run as well as an increased focus on everything the organization does in garrison and in the field. All of us junior leaders expected this new commander and first sergeant to be on our tanks, in our rooms in the barracks, and in our routine business more than usual. To my surprise, they weren't. Although they did pay attention to detail, they empowered us with responsibility, and they steered our company through an extraordinary transformation over the next two years.

Captain Bessler and First Sergeant Pastine made an outstanding command team. Their rapport—how they communicated with each other, how they respected each others' roles, and how those roles worked together to move the organization forward—established the foundation for command-team relationships for the rest of my career. The lessons I learned from them took me to a new level of maturity and leadership, and none of my success in the Army would have been possible without the training and example of these two great leaders at a key point in my career.

As part of the leadership transition, I was moved from being the gunner on the 1st Platoon leader's tank to gunner on the company commander's tank. At the time, the company also transitioned from the standard M60A1 tank to the M60A1 RISE/Passive tank.[7] I had never seen a new tank before. It felt like a new car—everything worked perfectly. It even smelled new. Fresh green paint on the outside, white interior, and brand-new tools and equipment—this was Christmas for a tank crewman.

Being assigned to the company commander's tank meant I had a great driver who had been handpicked for the honor of driving the commander's tank. My loader, new to the unit, was a sharp-looking Soldier from Puerto Rico with a four-year college degree. Captain Bessler focused on making the company the best in gunnery, the best in maneuver exercises, and the best in preparation for executing and fighting our wartime mission in our assigned general defensive positions (GDP) close to the East/West German border. Meanwhile, First Sergeant Pastine focused on building a disciplined and cohesive team, ensuring all company systems were up-to-date and ready for any surprise inspections, getting Soldiers to their required professional development schools early in their careers, and having clean and well-organized barracks.

Looking at these focus areas of both the commander and the first sergeant, one could see where their priorities might conflict, but they found the right balance to meet both sets

---

[7] The RISE (reliability improvement of selected equipment) program featured a new engine configuration. Passive IR sights allowed for visibility during the night without the need for infrared illumination.

of priorities. Preparing for big events like a tank gunnery or a maneuver density requires a lot of training and preparation. After that, there wasn't much free time left on the schedule, but the first sergeant worked his magic: every week, unless we were deployed, he carved out one hour of professional development time for the NCOs.

Every Thursday at 0530, Pastine taught us the standards for everything in and associated with the company. These early-morning mentoring sessions, held in the basement of our barracks building, were mandatory for all NCOs and all specialists pending promotion to sergeant. Every Thursday morning before sunrise, our place of duty was sitting in the front row of chairs. The second row of chairs was for the staff sergeants, and the platoon sergeants stood in the back of the classroom with their cups of coffee.

First Sergeant Pastine taught every single class for the next two years. While some leaders might resist investing the time and effort necessary to learn the subject areas to a level where they could teach it, Pastine relished in this proficiency. He devoted the time and effort to become a subject-matter expert, so he never needed to use an outside specialist to teach a class. I believe he did this not because he didn't trust outside speakers or teachers to accurately present the material, but because he wanted his own priorities and focus to influence the growth of the organization.

I was quiet and shy, so I looked to the first sergeant as an example of someone with the right focus and confidence. Pastine taught standards for the barracks, including each Soldier's room, wall locker, and bed. He taught us the standards of cleanliness for the common area latrines, shower rooms, hallways, and stairs. He taught all the tactical, company-level classes, from company fire plans and tactical assembly area operations to casualty evacuation, resupply operations, and more. He wasn't just a masterful leader—he was a conscientious, meticulous teacher.

Every NCO was assigned at least one additional duty in the company. I served as the drug and alcohol education specialist, managing the company's monthly random drug testing. I also served as the map custodian for the 22 sealed canisters—one for each vehicle in the company—that contained an accurate, updated set of the wartime maps. Neither assignment was difficult once learned, but they did require routine attention to maintain currency and meet the standards of an established checklist used by outside inspectors.

Other areas in the company included dental and physical readiness. Every Soldier required an annual physical and dental exam. An NCO was assigned to check in with the dental and medical clinics and then to notify and remind Soldiers and their leaders of their scheduled appointments. Field sanitation kits, hometown news releases, mailboxes, furniture inventory, and a host of other additional duties were all delegated to individual NCOs to manage. Each of these additional duty areas had clear, established standards that our first sergeant expected us to meet when he inspected them.

Learning those standards was certainly valuable, but to us young grasshoppers, the greatest benefit of those Thursday morning sessions was learning what we needed to be, know, and do as a leader. Pastine taught us that the science behind leadership is perishable and must be revisited and taught periodically, even when there are no leadership transitions. Going over these subject areas again and again was key to the first sergeant's success. He was also a believer in teaching two levels down, to both the tank commander and tank gunner levels of the organization. While most of us in those positions were sergeants and staff sergeants, his teachings set the tone and focus for everyone throughout the organization—mechanics and supply and training room personnel joined us tankers in these classes, and we were all taught his standard expectations for an NCO.

Throughout my career, I followed Pastine's example of regular review. I found that sometimes people don't mentally absorb a lesson the first time, so revisiting it later usually leads to deeper understanding and meaning. I myself gained new insights as I prepared these lessons, and I often found a new real-life situation to illustrate a teaching point. From watching my first sergeant, I learned that when subject material is important to an organization, it must be repeated and discussed, over and over again, to drive the understanding down to the lowest levels.

This repetition is the key to embedding and embracing the mission of an organization. For example, in January 2000 (20-plus years later), as the newly assigned division CSM in the 1st Armored Division, I supported my division commander Maj. Gen. George W. Casey Jr. using this technique to build a division team preparing for deployment to Kosovo. I found by repeating our division's mission statement and purpose in every unit visit or professional development forum, all Soldiers at every level of our 14,000-Soldier organization knew not only the division's mission, but now embraced the mission and purpose as their own. Even to this day I can recite the Iron Soldier mission statement.

For months, I had the perception that Pastine was running the company and setting the focus for the NCO corps, but looking back, I realize he and his platoon sergeants were working together for our professional development. The entire senior leadership team all looked for opportunities to improve and strive for excellence in all areas; they were completely in sync in their focus and effort.

Pastine was an exceptional teacher and example of the Army's 11 principles of leadership, valuable reminders of what I was doing right and what I needed to improve. They became my yardstick for measuring and understanding my strengths and weaknesses. Using these principles throughout my career allowed me to change my outlook on leadership and life in general.

I learned these principles in alphabetical order, so I could recite them from memory while appearing before my promotion boards.

1. Be technically proficient
This was the late 1970s, and Pastine and my platoon sergeants were Vietnam-era veterans. For those who fought in combat, being technically proficient often meant the difference between life and death for every Soldier in your unit. For the rest of my military career, I used and implemented the programs Pastine and Bessler put together to develop competency on the tanks and to help us become experts in our profession.

I was fortunate to have technically proficient leaders early in my career, both at Fort Hood and in Germany. These leaders were subject-matter experts in their profession; they taught not only the procedural steps and processes but also the scientific explanations of *why*. Having a deeper knowledge of the subject material taught me adaptive reasoning for specific situations: I could diagnose problems before they became disasters and take immediate action to remedy the issue.

2. Build the team

The leader of any group is responsible for building the team. Whether it is an officer working with his NCO counterpart, or a sergeant and two or three of her Soldiers, the leader is the one responsible for the cohesiveness of the organization. Team building is one of the hardest tasks for any leader, especially a new first-line supervisor or manager who is joining an established team.

Although a new leader doesn't necessarily need to "prove" himself when joining an established team, he does need to command respect and trust—and needs to show the same to his troops. Treating people with dignity and respect means talking to your subordinates as you would want a senior leader to talk to you. It means addressing Soldiers by their ranks and last names, eliminating profanity from your vocabulary, and reserving criticism for private, one-on-one counseling. These fundamentals are essential for building a cohesive, loyal, and trustworthy team.

While not originally codified as they are today, Army values have been with us since our humble beginnings as a nation. Every Soldier should know these values by heart, as well as their importance to the success of the team:

Loyalty is being devoted and faithful to your nation, your Army, and your fellow Soldiers.

Duty is performing what is expected and remaining loyal to the ones you serve, no matter where you are stationed or deployed.

Respect is treating fellow Soldiers, people, and family with dignity. Being respectful in and out of uniform takes dedication and practice.

Selfless service is putting the Army and your Soldiers before self. The mission of the organization must outweigh the personal needs or wants of the individual. When I became a leader, I decided I would always be present when I had Soldiers working. Even when I was a first sergeant, if I had troops in the motor pool working after hours, I worked right alongside them. People talk about balance between work, family, and life, but balance is not always 50/50—sometimes it's 90/10, depending on the mission. During tough times, a leader's presence reassures Soldiers that their time and effort is valued.

Honor is serving the American people and our elected officials in such a way that the Army and every organization within the institution can be trusted without question. A Soldier demonstrates honor when he does what is right all the time, in and out of uniform, even when superiors are not present.

Integrity is honesty and reliability—the glue that keeps an organization working together in harmony. Integrity will always be expected of those in uniform; for a leader, it's imperative.

Personal courage is doing the tough tasks that are expected of a Soldier and a leader: overcoming fear, adversity, or danger to stand up for or take action.

3. Develop a sense of responsibility in your personnel

I often ask newly assigned platoon leaders or sergeants to tell me their most important task or mission in the platoon. The most frequent responses are generally focused on training for the mission, readiness of the organization, or the health and welfare of the troops.

These things are all important, but none can be accomplished without leaders.

Growing sergeants within the platoon—that next generation of leaders, the "bench" of the organization—is critical if we want to accomplish priority tasks and missions. Putting someone in charge of a job or associated standards is one of the best ways to empower a subordinate and help him develop a sense of responsibility, a crucial first step for becoming a leader. Growing future leaders is part of growing future readiness in an organization.

4. Ensure the task is understood, supervised, and accomplished
One of a young leader's biggest mistakes is failing to make sure that a subordinate understands an assigned task. Providing the understanding, expectation, and final end state of any job creates the same vision for everyone involved. The more thoroughly someone understands the purpose of the task, the more thorough the result will be.

Ultimately, however, the responsibility for a task falls on the leader assigned to supervise those charged with accomplishing the task. Make sure your subordinates know the expected standards.

5. Employ your team according to its capabilities
As a first-line supervisor, manager, or leader, you must know your employees. For those who have never had a psychology class, this principle means knowing what motivates your employees and what they value in life—in other words, knowing what makes your team members "tick." Leadership is the art of influencing others to accomplish the mission with the minimal amount of resources.

More importantly, you must learn each individual's capabilities as well as their physical, mental, and intellectual limitations. Look at the type of operations you expect to conduct, and consider the physical demands and the stressors that might push someone over the edge. You must consider human dynamics in all operational planning. Failure within a small or large organization breeds distrust and fractures the team. When planning and executing their missions, leaders must factor in individual tolerances and resources to ensure their unit succeeds.

6. Keep your personnel informed
To build a successful team, keep your subordinates aware and knowledgeable of their surroundings. An effective leader will take responsibility to provide feedback from senior leadership, while also managing how information gets passed from the lowest levels of the organization back to the top. Every Soldier in the organization has a voice. This upward and downward flow of information serves to "level the bubble" of information across the organization by keeping people informed.

As a young leader, I took notes on anything senior leadership said and passed these words verbatim back to my Soldiers. This habit empowered me with the knowledge and wisdom to teach, educate, and inform my subordinates. I also felt like an advocate for my Soldiers when I could take their ideas and recommendations back to my leadership and help improve the organization from the bottom up. Sharing notes and ideas is a team-building effort that reduces stress for the Soldier in the organization and at home.

7. Know yourself and seek improvement

I was fortunate to have learned from great NCOs and officers throughout my Army career. Every one of them is as different as their fingerprints, and each has their own strengths and weaknesses. Be aware of what yours are. One of my strengths is the ability to recognize good and bad—I owe that trait to my mother and the time we spent together in church. One of my lifelong weaknesses is being an introvert. Even to this day, I have to intentionally push myself to be out in front of people, greeting them and initiating a conversation. It's a struggle to not revert to being the quiet shy kid hiding in the back of the classroom.

Knowing where I needed to improve, and copying the good traits of inspiring leaders, helped me grow taller and broader in my abilities to deal with complex situations and individuals. For many years, I dismissed the need for college education, as I believed I had greater common sense and judgment than many leaders with college degrees. In time, I understood the value of educational development beyond high school, including the ability to develop critical thinking and analytical skills. Once I learned the difference between common sense and critical thinking skills, I actively sought opportunities to improve this element of my leadership ability.

8. Know your personnel and look out for their well-being

Looking out for your people and taking care of them and their families applies to their personal and private lives. As an introvert, I never felt comfortable "prying" into a person's private life. What I knew about a person was whatever they willingly shared. From observing First Sergeant Pastine, however, I learned the value of keeping a leader's notebook with basic information about each of my Soldiers. This private notebook became part of my formal monthly counseling sessions as well as casual conversations. But the most important result of this notebook was establishing trust between my Soldiers and me. If they shared something confidential or personal, I safeguarded that information and established a rapport with my Soldiers as a trusted agent.

In addition to being familiar with your personnel's private and family lives, be aware of their careers. Professional development schools and courses are important for progression and assignment opportunities. Helping your Soldiers prepare is just as important— partnering a future student with an alumnus of the institution can go far to successfully getting the student into school and setting them up for greater success at graduation. Preparation for attending schools and courses sometimes involved carving out a few hours each day for a week, but the more technical and physical courses such as master gunner or ranger required weeks and sometimes months to mentally and physically prepare.

I watched our organization prepare and send many Soldiers to professional development schools and specialty courses where they excelled, often graduating with honors. Not only did these preliminary steps set the individual up for success, the payback for the organization was many times more as these successful students brought greater knowledge and expertise back to the unit. I have no doubt the success of the company over the two years when Pastine was the first sergeant could be attributed to this simple process of preparing students for school.

In addition to the military and civilian schools and courses available for Soldiers, also consider assignment opportunities. I watched Pastine rotate his driver back into the organization each year, which left a big impression on me. Soldiers assigned to driver positions are generally detailed into these positions to work outside the occupational specialty. Keeping Soldiers too long in one type of job without rotating them into career-enhancing or broadening positions makes them less competitive for promotion. Leaders who can see the future potential of individuals and their contributions to the institution will make the sacrifices necessary to send their people to school or rotate detail-assigned Soldiers back into the unit. The future of the Army relies on the advancement of our best Soldiers.

9. Lead by example

For me, this was the most important of the 11 principles; it served as an all-encompassing umbrella that covered all the other principles. I had to be the example to my Soldiers and my organization in all aspects of leadership, including my daily appearance, how I wore my uniform, my grooming standards, and how I spoke to my seniors, peers, and subordinates, as well as my technical and tactical knowledge of my career field, the organization, and the Army. Today, this area has greatly expanded to include everything from how leaders portray themselves on social media to how they dress and present themselves in their off-duty time and how they treat their families.

Leaders have a deeper obligation to those who choose to follow in their footsteps. Those who serve in positions of increased responsibility must embrace the great charge we have for influencing members of the workforce and society. As role models, we have a duty to demonstrate to others, especially those with less experience and maturity, what the right standards look like.

10. Make sound and timely decisions

I learned to make decisions as early as possible to allow my Soldiers plenty of time to prepare for future missions. Pastine believed that an 80-percent solution now is better than a 100-percent solution hours, days, or weeks into the future. No one likes to jump through hoops at the last minute. Even the troop-leading procedures I learned in the Basic Leadership Course were designed to get an organization preparing for a mission well in advance of receiving a formal mission statement or operations order.

Making sound decisions is both an art and a science. The science part is the easiest to understand: accounting for the available resources and the number of man-hours needed to complete a task gives first-line leaders a quick time estimate based on their organization's capability. The "art" of making a decision is based on the experience and wisdom gained from performing the task multiple times under a wide range of conditions—i.e., at night or during periods of limited visibility, reduced manning levels, mission requirements, enemy forces, terrain and weather, time, etc.

Changing from a day operation to a night one, for example, adds a degree of complexity to a task. Losing a subordinate leader or key player in the operation requires a leader to think about alternate solutions. These changes in the task create added experience and insight that will strengthen a leader's ability to make better, quicker decisions in the future.

11. Seek and take responsibility for your actions

One of the key attributes senior leaders look for in young potential leaders is the desire to take on responsibility. Those who seek and demonstrate responsibility for their behavior, tools, equipment, and personal lives have the characteristics of a hard worker. A leader who owns his stewardship and is accountable for subordinates and the work they perform can be trusted to represent the organization.

Taking responsibility as a leader means being answerable for all your organization does or does not accomplish. It's not enough to make sure that routine tasks are carried out on schedule—look for ways to improve effectiveness within your part of the organization. A successful NCO is both a manager and a leader: he must oversee and ensure the completion of routine tasks, but he should also look for opportunities to make the Soldiers and their small unit better able to support the larger organization's vision and mission.

Pastine's lessons over the two years he was first sergeant ran the gamut from big-picture leadership all the way down to the minute details of cleaning standards for the common areas, latrines, laundry room, TV room, day room, hallway, and stairwell in the barracks. He personally walked through all the areas after the morning physical training and personal hygiene to inspect our cleaning, while sergeants and staff sergeants stood by their teams and squads and respective areas of responsibility. Pastine had a sharp eye and knew where to find soap scum and hard water spots. By pointing out the not-so-obvious ugly things lurking in the showers, sinks, urinals, and commodes, he taught us young leaders to pay attention to detail.

I actually looked forward to the first sergeant's walk-through inspection every morning, because we were prepared. I felt such pride in my Soldiers as we finished our chores each morning and waited for Pastine to examine our work.

These morning cleanup chores taught us that if you are going to do a job, do it right the first time. We also learned that Soldiers and subordinate leaders are only as good as their senior leaders. Pastine's presence in the barracks set the tone for what was important and ensured that all of us were there. Having all the Soldiers and all the leadership present at one time during the day promoted morale and esprit de corps, because everyone contributed to the completion of a successful mission. This effort also ensured we lived in a clean, healthy, neat facility that even our mothers would be proud to show off. ■

Lesson Six: Be Ready for Opportunities

The successful warrior is the average man, with laser-like focus.
~ Bruce Lee

While First Sergeant Pastine worked to build a knowledgeable NCO corps through his Thursday morning lessons and the daily inspections of the barracks, Captain Bessler worked on making us the most lethal tank company in the Army. The focus on tank gunnery at Fort Hood had been important, but nothing like it was in Germany. At Fort Hood, I had learned how to read a map, navigate, maneuver as part of a larger formation, and maintain a tank and its weapons in the field; in Germany, we learned how to defend Western Europe from invasion.

With the real-world threat of the Soviet Union during the Cold War, being stationed on or near the East/West German border brought a sense of urgency to our mission. Even the Soldiers' families felt the threat of the Soviet Union. Every time the siren alert blared, usually in the early morning hours, we prayed it was an exercise and not the real thing. These alerts brought all the leadership and Soldiers running to report to their units, draw individual and crew-served weapons, activate communication security equipment, and prepare vehicles to move to pre-assigned assembly areas in the forests surrounding the nearby villages.

As the gunner on the commander's tank and later as a tank commander, I had to be one of the first vehicles ready to shoot, move, and communicate at a moment's notice. The "move" and "communicate" pieces of my job were second nature after serving on the battalion commander's tank at Fort Hood, but the "shoot" part was suspect, especially after I had failed six months earlier as a last-minute tank commander in 1st Platoon. At that time, tank gunnery focused on the accuracy of the tank's fire-control system and crew duties, how to engage when fighting outnumbered, the speed of engaging multiple targets, the engagement of the most lethal targets first, and engaging multiple targets with the correct type of ammunition or weapon to achieve target destruction. I found the art and science of tank gunnery fascinating.

I learned a great lesson from Captain Bessler during the two years we were together in B Company. Like Pastine, the commander focused on teaching the fundamentals. His tank gunnery skills test measured whether members of the firing crews could perform all 18 tasks safely and to standard; these tests included clearing, disassembling, inspecting, assembling, function checking, and loading the crew-served weapons and 105mm main gun. But that wasn't all: Bessler required tank commanders and their gunners to read the tank gunnery manual from cover to cover. There are jokes about Soldiers in the Army not studying or reading their own doctrine—but not in Bessler's organization. His initiative forced people into the gunnery and operator manuals to study, learn, and apply their skills in exercises.

Bessler's attention to detail was evident from the first time he and I performed dry fire drills. These were the days before computer simulation, so crew drills were learned on the tank, usually in motor pools or local training areas. We spent hours firing simulated bullets at the windows and doors of the barracks behind our motor pool. Giving fire commands, adjusting fire, engaging multiple targets, and participating in simultaneous engagements with the main gun and machine guns were the building blocks of our drill rehearsals. When we mastered one task, we would transition to a more challenging one, continuously rehearsing each type of scenario to mentally prepare ourselves to fight the Soviet Army or the "Plywoodian" Army of the tank crew qualification range.

Right up until qualifications at Grafenwöhr, I practiced with Bessler to master the task of engaging targets on the move and from the short halt. All our work in the motor pool and in the training area paid off: with a distinguishing run and a nearly perfect score, we were the second-best tank crew in the battalion, and the *Stars and Stripes* newspaper reported that one crew out of the 3rd Platoon fired "the perfect" run. Coming in second out of 54 tanks was a great feeling. The experience preparing for this gunnery qualification completely changed the way I approached training.

I learned that attention to detail in any task is the key to attaining excellence. I liken what we did to a professional athlete who continuously trains and develops muscles, dexterity, and timing to achieve excellence in their sport. With a tank, it's more complicated, because you have four crew members who must master their individual skills before applying those skills collectively as a crew; a tank is a robot where all four crew members are the brain and must think alike and know what each other is doing. I learned that the secret to training individuals and developing lethal tank crews is to focus on details.

The next summer, I learned that unexpected events often require leaders to advance Soldiers into positions of increased responsibility, and you have to be ready to grasp those opportunities at a moment's notice. The commander of another tank was reassigned, and I was selected as his replacement—just 45 days before Level 1 gunnery. Thoughts of failure swept through my mind as I remembered the tank crew qualification one year earlier. While I had become a very successful gunner since then, I had been training for four months with a great tank commander and crew.

Not only did I have just a few weeks to prepare, I now had an inexperienced crew and an inoperable tank. The new tank I left behind was perfect in every way—clean, organized, maintained, and fully operational. The tank I moved onto was the complete opposite; in fact, the engine was literally sitting on the ground, waiting for a replacement engine-oil cooler.

My new gunner, a specialist, had been serving as the driver for the first sergeant. My loader, a newly assigned private first class, had just arrived from Fort Hood, and my driver was a specialist 5 who was due to leave the Army in four months. I'll admit there was a moment—looking at the short preparation period, inexperienced crew, and broken tank— when I wanted to say "no thanks" to becoming a tank commander.

Fortunately, my new platoon sergeant was Sgt. 1st Class Hugh (Skip) Ashton, a former tank commander in Vietnam. He had received a Purple Heart for injuries sustained when his tank was literally shot out from under him, killing his gunner and loader. Ashton was a tactical expert on the employment of tanks on the battlefield and often served as a platoon leader in the absence of a lieutenant. Captain Bessler always took Ashton—a giant in the armor community—with him to the GDP briefings on the border with the battalion and brigade leadership. Years later, I was fortunate to work with and grow under Ashton in my future assignments with the Weapons Department, U.S. Army Armor School at Fort Knox, Kentucky. Ashton's mentorship and friendship continue to this day.

For the next 45 days, Ashton was my constant coach and mentor. He worked with me to ensure I could set up the tank's fire-control system to engage targets accurately, and reminded me what needed to be done prior to each phase of the gunnery density. Following Bessler's example, the crew and I spent hours each day, during lunch and after duty in the motor pool, rehearsing tank gunnery engagements. Slowly, we grasped the fundamentals and grew our mastery of gunnery skills to include adjusting fire, engaging targets with the telescope, and engaging multiple targets using multiple tank weapons. Our loader served as a timekeeper to measure our progress and mastery of each specific skill.

The big day finally arrived: after failing as a tank commander the year before, I'd been given another opportunity. We got there early in the morning to boresight[8] and prepare the tank and weapon systems. Everything checked out, and we were ready. I cannot adequately describe how nervous I was—I just couldn't fail again. I was also determined to not let anyone know how nervous and scared I felt.

It was July 1979, and the Army had just completed building the infamous Range 10 (now Range 117) within the previous six months. We had all heard stories about the difficulty of this range and the low number of tanks that passed qualification. The range had four different lanes and 36 different potential target scenarios, so there was no way to "G2" or memorize the target engagements or locations, like some had done on previous ranges. We were not told which lane to occupy until we pulled from the motor pool onto the range complex—this was truly like fighting your tank in a combat situation.

Of the 17 tanks in the company, I was the second tank in the platoon and the fifth tank to shoot. Because of the challenges and complexity of this new range and my inexperience as a tank commander, Ashton received permission to ride on top of the tank turret to coach me if needed. The range was set for us to shoot seven engagements during the day and return to shoot three engagements at night.

We attacked the range, and everything went exactly as rehearsed. The tank was deadly accurate, and both the .50-cal and 7.62-cal machine guns fired perfectly. A slight breeze helped clear the dust after each main gun round fired. We even hit a bonus target that had been added to the range scenario, a small turret silhouette nailed to the top of a 40-foot telephone pole and placed at the back of the range. We had 15 seconds maximum to engage the target when designated by the control tower, nearly a 3,000-meter shot from the baseline. We aced it dead center.

The after-action review was great. All the hard work and hundreds of hours of training over the past two months had paid off. I cautioned the crew that we still had to shoot our three-night engagements, and we needed to continue the precedent we'd established with our day run. The night runs turned out perfectly as well, and we ended up the highest-scoring crew in the company that day. By the end of the week, after all the other tanks in the battalion finished, we were the top tank.

---

[8] Pre-aligning or calibrating the main gun barrel to the aiming devices; making sure your sight is pointing where your gun is.

I was proud of the crew we'd built over the last 60 days, but that glory was short-lived. My gunner received assignment orders for Fort Hood, and my driver had chosen to leave the Army; both departed the unit in the following 60 days. With no new personnel in sight, I planned to make my loader the gunner, but Pastine pulled him to drive Captain Bessler's Jeep.

Ashton, my platoon sergeant, was promoted to first sergeant and reassigned to Fort Knox, and Sgt. 1st Class Richard (Dick) Morgan took his place. After spending a year working with Ashton, I never expected to see another platoon sergeant of his caliber and with his knowledge of the tank—but I was wrong. Morgan came into the unit as a former drill sergeant with a reputation for being tough and demanding, but he was a great trainer and he was fair. I learned much from him. As with Ashton, I worked for Morgan again in the Weapons Department, U.S. Army Armor School at Fort Knox, and as a first sergeant in the 11th Cavalry Regiment. Morgan became one of my greatest mentors throughout my career, and a lifelong friend.

The news of losing my loader left me feeling hopeless; all the time and work I had invested into these Soldiers was lost. I was now a one-man crew. Eventually I was assigned two rehabilitation Soldiers from other units to serve as my driver and loader, but that didn't help me feel much better—now I was getting all the misfits.

As Pastine made his rounds in the barracks the following morning, I could not keep silent. I questioned the loss of my loader to serve as the commander's driver. He patiently listened to my concerns about training and the investment of time to prepare my previous crew and how I was getting a raw deal.

"Sergeant Preston, that's your job," he said, looking me right in the eyes. "That's what you do."

As he turned and walked away, the reality of those words sank in and I learned yet another lesson. He was right. That's what sergeants do: they teach and grow future leaders. Sometimes it's not until you are hit in the face with the obvious that you truly understand something. I picked up my pride and determined to move forward and build a new crew, even with Soldiers who were rejected and cast out by others. *You can do this*, I thought, giving myself a pep talk. I still needed a gunner, though, and my search continued while I hoped a stellar new Soldier would arrive in the unit. Days and weeks passed, and I was running out of options. I needed a gunner to start training for the next gunnery density, which was now only about 90 days away.

To complicate things, Pastine selected me to appear before the Sergeant Morales Club board at brigade headquarters. The Morales Club was an initiative started by Gen. George S. Blanchard in the 1970s, when he was commanding general of U.S. Army Europe. The initiative was designed to recognize NCOs who epitomized "Sergeant Morales," the fictional name of a real-life sergeant who had set the standard for what an NCO should be, know, and do. As I was the tank commander of the top tank crew in the battalion, Pastine felt I deserved this honor.

I spent my nights and weekends studying manuals and Army regulations. The Sergeant Morales Club was a new initiative, and there were no members of this elite club in the company or even the battalion. I appeared before the selection board chaired by the brigade and four battalion CSMs. Looking back to that frightful day, by my own assessment I did okay, but nowhere close to the level of competence and polish I needed to move on to the next board in the Hanau military community.

I thought I was done, but Pastine immediately nominated me to appear before the next quarterly board. At this point, studying was a distraction; I was focused on training a new crew and I wanted to spend time with my family. Nevertheless, I complied and studied hard to prepare for the next brigade-level board. This time I was nominated to move forward to the community board.

The following month I made the short trip to Hanau, where I reported to the military community CSM and a panel of five senior brigade-level CSMs. To my chagrin, they asked tough situational questions that I was not prepared to answer. To make a long story short, I did not pass the board and returned to my unit. In the back of my mind, I was through with the Sergeant Morales Club. Or so I thought.

Meanwhile, I was still looking for a gunner, but the platoon had 20 Soldiers, exactly the number authorized for manning five tanks. There was one extra Soldier, a young man who had been the loader on the platoon leader's tank, but he had just received a Field Grade Article 15. Under the Uniform Code of Military Justice, company grade and senior commanders can initiate nonjudicial punishment for lesser forms of misconduct. In this young Soldier's case, he had violated a direct order from the battalion commander. His punishment was reduction to PV1 and 30 days in the CCF, the little prison in Hanau.

I was one of two NCOs detailed to take this Soldier's mail to him twice a week while he was serving his time. Each time I went to the jailhouse, this Soldier asked me if I had a gunner on my tank. He begged me to give him a chance as my gunner. At the end of his 30 days, the Soldier returned to the platoon, and I took him as my gunner. Now I truly had a crew of misfits—"the dirty trio," as my fellow NCOs hazed me periodically.

But with my newfound knowledge and confidence in my training abilities, I promised the crew that if they did exactly what I told them and exactly what I expected of them, they would have an opportunity to finish their service honorably. All three Soldiers nodded north and south, and we began preparing for the gunnery skills test and gunnery ranges in the fall of 1979.

This would-be tank crew was definitely hungry, eager for the opportunity to be good Soldiers and complete their enlistment in the Army—versus getting chaptered out of the force with a less-than-honorable or bad-conduct discharge. Admittedly, they were knuckleheads, and they tested my patience many times. I spent countless hours with them, individually and as a crew. But the reward of hard work and helping dreams come true is a lesson I will always remember. They were totally committed to meeting my demands and expectations to be good Soldiers, far beyond training for tank gunnery.

Morgan, my platoon sergeant, played a huge part in keeping me motivated and moving forward, and he taught me an important lesson: leaders should always be present, even if they're not technically needed. Within days of his arrival, we were selected to display our tank during the upcoming German-American Friendship Week. The crew and I took our tank down to the wash-rack area on a Saturday two weeks before the big community festival to put new camouflage paint on the outside and detail the white interior for public display. Morgan was there with my crew all weekend as we made sure our equipment and every detail was perfect.

Spending this time with our new platoon sergeant gave me an opportunity to learn more about the man and the Soldier, and he in turn learned much about us as a crew. I took this lesson and used it in every opportunity where I had subordinate leaders working into the evenings, weekends, and holidays—I stayed with them and was there to support them, their Soldiers, and their efforts.

My gunner set out to become "supernumerary" at guard mount 10 times, so he could be exempt from guard duty for 30 days. This term describes an extra member of the guard force who performs special duties or replaces a guard who becomes ill. With a total of ten guards reporting each day for duty around the motor pool, nine were needed to support the three shifts, leaving one as a standby replacement. This guard was selected based on personal appearance and responses to questions. As it turned out, my gunner was smart, dedicated, and committed to proving himself. He was the consummate professional at standing guard mount and soon attained 10 supernumeraries in a row.

One day, I was in my gunner's barracks room doing a final inspection of his uniform before guard duty when we had a surprise visitor. Command Sgt. Maj. Oscar L. Barker, the command sergeant major of V Corps, made an unexpected visit to the battalion, and naturally they brought him to B Company to see what a good-looking and well-kept barracks looked like. The door to my Soldier's room opened, and in walked Pastine and CSM Barker. After a quick introduction, Barker began questioning my gunner about the unit, training, tanks, and a host of other things. My gunner was sharp and answered every question accurately and professionally. Barker then turned to me.

"Sergeant Preston, are you a Morales Club member?" he asked.

I felt sick.

"No, Sergeant Major," I replied, embarrassed.

"First Sergeant, get this Soldier before the next Morales Club board," he instructed Pastine.

"Yes, Sergeant Major," Pastine answered, giving me a look.

For the next six months, I studied hard to prepare for the next series of Sergeant Morales Club boards. Third time's the charm: I finally passed the V Corps board and became a member of the prestigious Sergeant Morales Club.

I've told this story many times to illustrate the value of appearing—sometimes more than once—before these professional-development boards such as the Sergeant Morales Club, NCO and Soldier of the Month, quarter and year boards, promotion boards, and the Sergeant Audie Murphy Club. Each of these boards promotes self-development and intense study. Just as the gunnery board forces tank commanders and gunners to study their manuals, professional-development boards force young leaders to read doctrine and continue to learn throughout their careers. Today we refer to this as lifelong learning.

While becoming a member of the Sergeant Morales Club looked good on my efficiency report for promotion, it was the knowledge I gained from studying and appearing before those boards that made me far more competitive for promotion. An added bonus is that I could pass on this knowledge to my Soldiers to improve their potential for service and promotion.

My tank crew and I again spent hundreds of hours doing crew drills to fight outnumbered on the potential battlefield and the gunnery training range. This time we were assigned Range 20 (Range 112 today) for our Level II gunnery density. My crew and I were the second tank in the platoon to shoot. After firing six engagements in the day and four at night, my little band of misfits and I came in first in the battalion.

Once again, the principles I'd learned, the commitment and dedication of my leadership, and the command's belief in me to train a group of Soldiers unwanted by other units came together on a cold December day on a tank gunnery range. My battalion and brigade commanders knew each member of my crew, and when dignitaries visited the brigade and wanted to see a tank, it was my crew who hosted the visit and demonstrated what an Army tank looked like.

My "dirty trio" and I stayed together for another gunnery density in the summer of 1980, and then I was the one who left the crew to serve as the company master gunner. Since I knew the secret to training lethal crews, teaching other crews what I had learned was an exciting opportunity. And company master gunner was a great position to serve in prior to moving on to the next challenges the Army had for my family and me.

*   *   *   *   *

I've used these stories and many more to help young leaders who are striving to be their best understand the dedication and commitment they need to succeed. In my mind, I was a less-than-ordinary Soldier, a quiet, shy young man who was given the gift of knowledge. All I had to do was apply what I learned to turn an opportunity into a reality.

During my first five years in the Army, I was blessed to work under the highest caliber of leaders who served as role models and teachers. I applied the principles I learned as a sergeant in all the training I supervised when I became a platoon sergeant, then a first sergeant in a cavalry troop and a headquarters company, then as a battalion, brigade, division, and Corps command sergeant major preparing units for major training densities and combat, and finally as Sergeant Major of the Army, working to help prepare Soldiers who fought in two wars.

The fundamentals of leadership were soundly engrained in me when I left Germany in February 1981 on my way to Fort Knox, Kentucky, as a newly promoted staff sergeant and gunnery instructor. With a solid foundation of leadership under my belt and knowing what was expected of me as a NCO, I was confident I could be a success at any endeavor if I followed the principles I have described.

The lessons I learned during that first year in Germany set me up for success as a young leader and senior leader for the rest of my Army career. Knowing what it takes to attain excellence was the greatest lesson I could have learned, and I continue to use what I learned to this day: attention to detail, rehearsals and practicing, performing a task in a wide variety of conditions, and leaving no stone unturned. Again, there is no substitute for experience and opportunities to perform the task and fail, and fail, and fail again until there is no failure.

Effective, powerful leadership is an art that is most effectively learned and practiced through hands-on experience. One of the most important things I learned as a leader is that there is no one way to handle any given situation. Each situation is as unique as the individuals involved, and effective leaders will consider the personalities involved as they work and serve and lead. Over the years, I had the opportunity to work with thousands of individuals, and hopefully helped influence, mold, and change in positive ways.

As I reflect back on those early days in uniform, it's clear that the leaders I served and the life-long friendships I will forever cherish are the reasons I made the Army my career. The leadership lessons I learned in my first five years in the Army set the foundation and charted the azimuth for who I would become in 36 years.

I can honestly say that if I had the chance to do it all over again, I would. While I made many mistakes over the years, they taught me lessons that made me a better person and a better leader. I am a product of the thousands of Soldiers I served and the hundreds of leaders who influenced me. This is their story as well as mine. ■

Lesson Seven: Transitioning

You must expect great things of yourself before you can do them.
~ Michael Jordan

Throughout my Army career, I knew at some point I would leave military service and return to the civilian workforce. Like most young people today who volunteer to serve in our military services, I intended to complete my four-year service obligation and use my earned Montgomery G.I. Bill to attend college. I wanted to become an architect—never in my wildest dreams did I think I would spend 36 years in uniform.

Over the years there were decision points, crossroads at which my family and I had to choose whether to stay or go. Knowing I would leave military service at some point, I made a conscientious effort to have a skill I could use in the civilian workforce. When I was a young private first class in my first assignment at Fort Hood, Texas, all the Soldiers in my unit were encouraged to take Army correspondence courses to start accumulating points for promotion to sergeant and staff sergeant. While these promotions were far into the future, I listened to my leadership and signed up for my first course, 63B Light-Wheel Vehicle Mechanic. I always wanted to understand automotive mechanics, and learning how to work on M151A1 Jeeps and medium trucks was an opportunity to gain promotion points, learn how to perform maintenance and repairs on the family car, and develop a skill I could use anywhere in the civilian workforce. My experiences with the course were positive and ultimately led me to pursue a technical degree from the International Correspondence Schools in automotive technology.

I use these experiences to coach the next generation of Soldiers and leaders to plan for transition from the Army. Whether you stay for four years or 40, everyone who serves in the military must someday transition to a civilian job and community. How well you develop skills and options will determine your transition's success and smoothness. While I didn't want to become a full-time automotive or diesel mechanic, at least I had the knowledge and skills to give me a head start going into a competitive civilian workforce.

I have three recommendations for transitioning from the military to the private sector:

1. Never stop learning
You can't start preparing for the move to the civilian sector just a few months before—always stay ready to transition by continually learning and improving. In the Army, we want Soldiers to pursue lifelong learning opportunities not only in their military profession but also in leadership training. I have taken hundreds of hours of Army correspondence courses, dozens and dozens of military and civilian classes ranging from public speaking to computer technology, and ultimately received a master's degree in business.

The philosophy behind the Army's Soldier for Life program is based on a three-phase life cycle in the Army: start strong, serve strong, and continue strong. Recruits joining the all-volunteer force today *start strong*: they are the "crème" of our society, and most come from the top 29 percent of high school graduates. Initial entry training embodies the seven Army values and our warrior ethos. Soldiers *serve strong* throughout their career by learning new skills, working with a diverse workforce, and attending professional development schools throughout their military service.

Helping Soldiers *continue strong* after military life is a high priority for the Army. Starting a year before retirement, transitioning courses and individual counselors are available. Learning new skills and pursuing college or technical school credentials throughout your military career makes this transition process even easier. Take advantage of tuition assistance. Most colleges offer online and distant-learning degree plans for those who have full-time jobs and families.

You'll find that your credentials are communicated differently when you enter the civilian workforce. In the Army, we wear our credentials on our uniform. Seeing the rank of a private, first sergeant, captain, or lieutenant colonel immediately telegraphs a certain set of skills, experiences, and expectations. In the civilian workforce, you have to look at a résumé to see credentials and degrees. The Army will soon have some type of certification for every occupational specialty. Acquire all the credentials you can while in uniform; you'll need those degrees, licenses, certifications, and qualifications when you transition.

2. Get medical checkups

Don't leave the Army with any deferred maintenance on your body. Serving in the military is physically challenging, and while some career fields are more demanding than others, all Soldiers are subject to periodic injuries. See a doctor and document any care or medication. Having an accurate account of your injuries and treatment will make the transition from the Army to the Veterans Administration and civilian healthcare far easier.

While seeking medical attention may not be a major concern if you're young, injuries may plague you with problems decades down the road. Don't wait until your last days in uniform before you go to the doctor to get a list of things fixed. You may find yourself transitioning into the Veterans Administration system for care that should have been accomplished while in uniform.

3. Establish priorities early

Decide early on what your priorities are. Where do you want to live? If you're open to a range of locations or income, you have more options. I was taught that if you enjoy what you're doing, you won't work another day in your life. For me, it was not about the money—I sought opportunities to have fun and make a difference.

* * * * *

On August 30, 2011, I retired from the Army. I demoted myself to private and returned to my hometown to get the farm up and running again. Going home gave Karen and me the opportunity to take care of our parents and give something back to them for all the opportunities they gave us early in life. I enjoyed being the groundskeeper and handyman for my parents again and for my father-in-law, who retired from the Army Reserve after more than 26 years.

Since retiring, I have served on the board of three organizations: Homes for Our Troops, a nonprofit organization that builds and donates specially adapted custom homes for severely injured post-9/11 veterans; the United Services Organization (USO), a group that takes care of our service members around the world; and the Institute for Veterans and Military Families at Syracuse University in New York. Syracuse created an entrepreneur boot camp to teach transitioning veterans how to start and run a business; their efforts have since expanded to include programs focused on women, and they partner with many colleges and universities across the country. All three organizations provide valued service to our military families, and it is an honor to give them my time and talents.

In February 2012, I volunteered to serve as the interim president for Homes for Our Troops. This small organization in Taunton, Massachusetts, offers a home to our most seriously wounded veterans. In April 2012, I was appointed co-chair of the chief of staff of the Army's Retired Soldier Council. This 14-member council is comprised of seven officers and seven enlisted retired Soldiers representing the retiree councils at posts, camps, and stations across the country and overseas. The council represents more than one million retired Soldiers and surviving spouses.

In May 2013, Gen. Gordon Sullivan (retired), a former chief of staff of the Army and then-president of the Association of the United States Army (AUSA), asked me to serve as the director of NCO and Soldier programs. Principally an educational association, AUSA was formed in 1950. Gen. Carter Ham (retired) became General Sullivan's successor in 2016, and I continue to serve the AUSA as the vice president for NCO and Soldier programs. The organization's mission is to serve as a voice for the Army and support for our Soldiers, families, and Army civilians across all three components of the Army. I'm grateful for the opportunity to continue serving, teaching, and mentoring with a great team, now in a new capacity.

In June 2014, I was sworn in as a member of the Defense Advisory Committee on Women in the Services (DACOWITS). Established in 1951, this committee is one of the oldest federal advisory boards still in existence. These last three years have been monumental for our military services, as the remaining occupational career fields closed to women are now open. DACOWITS studies a series of topics each year and presents an annual report to the Secretary of Defense. I serve as the vice chair of the committee with Gen. Janet Wolfenbarger (retired).

I also serve on the board of directors for the Army Historical Foundation, which raises funds for the National Museum of the United States Army at Fort Belvoir, Virginia. This facility will tell the history of our nation through the eyes of our Soldiers who have served since the Army's beginnings in 1636.

I volunteer as a director for the Armor and Cavalry Heritage Foundation, and support their efforts to build an armor museum at Fort Benning, Georgia. With the move of the Armor School to Fort Benning and the creation of the Maneuver Center of Excellence as part of the 2005 Base Realignment and Closure Commission, no facility currently showcases the priceless historical artifacts that were displayed at Fort Knox as part of the Patton Museum. The new museum will tell our Soldiers' stories for future generations.

I serve as a military advisor for First Command Financial. As a customer for many years, I owe much to this institution for teaching me the importance of saving for the future and starting early in life. Financial education for our service members and their families is critical for their long-term financial health and quality of life, especially with the implementation of the new DOD blended-retirement system.

I also serve as the honorary command sergeant major of the 11th Armored Cavalry Regiment based at Fort Irwin, California. I served in the 11th Cav during the fall of the Iron Curtain and Desert Storm/Operation Positive Force, and I owe much of who I am today to my experiences during that time. In their present role at our National Training Center, today's Blackhorse troopers continue the legacy of the regiment that began during the horse cavalry period of our nation.

I still have my hobby farm with a couple of horses, several antique tractors, and the equipment to produce my own hay each summer. I enjoy turning wrenches and working on the collection of old tractors and Ford trucks I've accumulated over the years. Along with hay, I planted potatoes and tomatoes this year. I have been volunteering to serve as the treasurer and a vestry member at Saint George's Episcopal Church in Mount Savage for the last five years.

In March 2017, I made a trip back to Afghanistan as part of Operation Proper Exit. Mr. Rick Kell, Command Sgt. Maj. Tom Capel (retired), Medal of Honor recipient Master Sgt. Leroy Petry (retired), five wounded Soldiers, and I traveled to Kuwait and Afghanistan for a week. Most of the wounded veterans only remembered an explosion before regaining consciousness in a hospital bed at Walter Reed Medical Center. Returning to their place of injury in theater, and then leaving under their own conditions, brought closure to a gap in their military service.

While I was serving as the president and CEO for Homes for Our Troops, I met the leadership of Veterans United Home Loans, the largest VA home-loan provider in the nation. VUHL and HFOT established a partnership with their sponsorship of specially adapted homes for severely wounded veterans. I continue to support VUHL as a military advisor in their quest for continued excellence and support to our veteran community. While there are many VA loan providers across the country, the personalized service and support to a veteran or service member and their family is excellent. They treat our military like they would want to be treated.

Having the opportunity to serve the Army in the Pentagon also gave me the opportunity to serve as part of the ultimate joint team. The senior enlisted advisors of our sister services were nothing short of exceptional and amazing warriors. I gained from their life experiences and relied on their wisdom and counsel on many occasions. In 2017, the six of us formed Summit Six LLC. Our goal beyond this book is to continue sharing our leadership experiences and lessons learned from our leaders and the millions of servicemen and women we stood alongside, shoulder to shoulder, over the course of more than 180 combined years of service.

Serving my community, both in Mount Savage and in the military, allows me to give back to the people who gave me a great start in life. My goal is to pass this attitude of lifelong service on to all veterans and retired service members—our youth need their leadership. ■

TOP: Sgt. Ken Preston receives an impact Army Commendation Medal from Maj. Gen. Walter F. Ulmer (far right) for the top tank crew score in the 3rd Armored Division at Grafenwöhr, Germany. Also in the photo is Capt. James E. Bessler (far left), commander B Company, 1-33 Armor battalion; and Col. Ernest E. Cross, commander 2nd Brigade, 3rd Armored Division.

ABOVE: V Corps command group in Iraq, April 2003. From left: Col. David Teeples, commander 3rd Armored Cavalry Regiment; Maj. Gen. David H. Petraeus, commander 101st Airborne Division; Maj. Gen. Ricardo Sanchez, commander 1st Armored Division; Lt. Gen. William S. Wallace, commander V Corps; Command Sgt. Maj. Kenneth Preston, V Corps CSM; Maj. Gen. Buford C. Blount, commander 3rd Infantry Division; Maj. Gen. Raymond T. Odierno, commander 4th Infantry Division.

RIGHT: Pvt. Kenneth Preston, July 1975, basic training.

TOP: Sgt. Ken Preston, tank commander B22 (M60A1 RISE/Passive tank nicknamed "Bandit 1"), Grafenwöhr, Germany, July 1979.
ABOVE: SMA Ken Preston talks to Soldiers of the 172nd Stryker Brigade Combat Team "Arctic Wolves" on his visit to Fort Wainwright, Alaska, July 13, 2005. The BCT was deploying in support of Operation Iraqi Freedom. U.S. Army photo by Staff Sgt. Reeba Critser.
LEFT: Commander in Chief's Inaugural Ball, National Building Museum, Jan. 20, 2009. President Barack Obama addresses the attendees surrounded by the senior enlisted advisors of the armed services. From left: SMA Ken Preston, SMMC Carlton Kent, MCPON Rick West, CMSAF Rodney McKinley, MCPOCG Charles "Skip" Bowen.

TOP: SMA Ken Preston talks with Soldiers of Joint Task Force Bravo, Soto Cano Air Base, Honduras, following a 4-mile morning run on Aug. 7, 2006. U.S. Army photo by Senior Airman Mike Meares.
ABOVE: SMA Ken Preston speaks to Soldiers of 1st Battalion, 502nd Infantry Regiment, after joining them for lunch July 8 at Forward Operating Base Mahmahdiyah. Photo by Specialist Kelly K. McDowell.
RIGHT: SMA Ken Preston talks to El Paso media during a break at the 2007 Nominative Command Sergeant Major conference at Fort Bliss, Texas.

ABOVE: Preston family photo, Jan. 2004. From left: Kenneth Jr., Karen, Ken, Valerie, and Michael.

TOP LEFT: PV2 Kenneth Preston, Karen Preston, and daughter, Valerie.

TOP RIGHT: SMA Ken Preston visits 10th Mountain Division Soldiers at Combat Outpost Korengal in Afghanistan.

ABOVE: The Command Sergeants Major of V Corps, Al-Faw Palace, Baghdad, Iraq, April 2003. Front row, left to right: CSM Chuck Fuss, 4th Infantry Division; CSM Charlie Thorpe, 82nd Airborne Division; CSM Ken Preston, V Corps; CSM Marvin Hill, 101st Airborne Division; CSM Roger Blackwood, 2nd Cavalry Regiment and later 1st Armored Division. Back row, left to right: CSM John Caldwell, 3rd Armored Cavalry Regiment; CSM Julian Kellman, 3rd Infantry Division; CSM Anthony Albain, 3rd Corps Support Command.

Micheal P. Barrett

Micheal P. Barrett served as the 17th Sergeant Major of the Marine Corps from June 9, 2011, to February 20, 2015. Born and raised in Youngstown, New York, he enlisted in the Marine Corps in March 1981 and underwent recruit training at 2nd Recruit Training Battalion, Delta Company at Marine Corps Recruit Depot, Parris Island, South Carolina. In November 1981, Barrett completed Infantry Training School at Camp Lejeune, North Carolina. He was ordered to 1st Battalion, 4th Marines in Twentynine Palms, California, for duty, and served in a variety of billets from grenadier to platoon sergeant.

Barrett was transferred in 1984 to inspector/instructor duty with 2nd Battalion, 25th Marines in New Rochelle, New York. His next assignment was to 3rd Battalion, 9th Marines, where he assumed the responsibilities as platoon sergeant of the Surveillance Target Acquisition Platoon. He forward deployed during the Gulf War with Task Force Papa Bear. In 1992, Barrett received orders to Drill Instructor School, Marine Corps Recruit Depot, San Diego. Upon completion, he was assigned to Company F, 2nd Recruit Training Battalion as a drill instructor, senior drill instructor, and chief drill instructor. In 1994, he was selected as the battalion drill master of 2nd Recruit Training Battalion.

In 1995, he was assigned to Scout Sniper Instructor School, Quantico, Virginia, as the chief instructor. In 1996, he was transferred to Marine Security Company, Camp David, for duties as the company gunnery sergeant and liaison to the United States Secret Service. Upon completion of his tour at Camp David, Barrett was transferred to 3rd Battalion, 4th Marines in Twentynine Palms, where he assumed the duties as India Company first sergeant from 1998 until 2000.

Barrett was assigned to Headquarters and Service Company in 2000; his tour culminated as the senior enlisted leader of Weapons Company from 2001 to 2002. Barrett was assigned to Recruiting Station Cleveland, Ohio, as the recruiting station sergeant major from 2002 through 2005. He was transferred to 2nd Battalion, 7th Marines in 2005, where he completed two combat deployments in support of Operation Iraqi Freedom in the Al Anbar Province, Iraq.

From 2007 to 2009, Barrett was assigned to Officer Candidates School in Quantico. He was selected as the 1st Marine Division sergeant major, assumed the duties as I Marine Expeditionary Force (Forward) sergeant major, and deployed to Operation Enduring Freedom in 2010. During this deployment, he also became the NATO Regional Command (Southwest) command sergeant major for Nimruz and Helmand Province, Afghanistan.

During his tenure as Sergeant Major of the Marine Corps, Barrett was a driving force in implementing the new Marine Combat Instructor Ribbon and new PME requirements for enlisted Marines.

Barrett is a graduate of 23 military schools and advanced courses, including the USMC Scout Sniper Instructor School and the U.S. Army Ranger School. His personal awards include the Navy Distinguished Service Medal, Legion of Merit, two Bronze Star Medals with combat "V" and gold star, Meritorious Service Medal with gold star, Navy Marine Corps Commendation Medal with combat "V" and three gold stars, Navy Marine Corps Achievement Medal with two gold stars, Combat Action Ribbon with gold star, and the Presidential Service Badge.

Barrett currently serves as a military advisor to Veterans United Home Loans (VUHL) and sits on the board of directors for the Marine Corps Scholarship Foundation and Young Marine National Foundation. He also volunteers for the Marine Corps League, a Veterans Service Organization, and the Armed Services YMCA.

He and his wife, Susan, have two sons and live in Bloomington, Indiana. ∎

Introduction: What Got Me Here

Some people spend an entire lifetime wondering if they've made a difference.
The Marines don't have that problem.

~ Ronald Reagan

I was 17 when I stood on the yellow footprints at Parris Island to begin my life as a Marine. It was hot and humid, we were soaked in sweat, and the squad bays smelled like feet, bad breath, and vomit. The recruit across from me had just tossed his cookies. Gnats moved freely in and out of my nostrils, and the drill instructor's spittle covered my face.

I was 51 when I stood at the Iwo Jima Marine War Memorial in Washington, D.C., to retire from my life as an active-duty Marine. Six inches of snow lay on the ground, and the temperature was minus-six degrees—so cold that the Marine Corps color bearer had a bloody nose and a bloodsicle on his chin. As I handed the sword of office to the commandant of the Marine Corps, I felt an incredible sense of pride, along with a sadness I had never experienced before.

And there you have it: my first and last days in uniform. In between were 12,408 days of hard work and constant learning. I never set out to be "successful"—I always thought it was normal to bust your ass and give it everything you've got. I wasn't the smartest guy to ever patrol or stalk across a battlefield, but I was trainable. There's no magic in what I did to become an effective leader in the Marine Corps—I just learned and mastered the basics to the letter, studied the Marine Corps philosophy intensely, memorized the creeds and code of conduct, applied them always, used checklists to make sure I didn't overlook anything, observed my leaders carefully, and relied on the examples and character of my fellow Marines. And at every opportunity, I shared what I learned. That's what got me here.

*   *   *   *   *

Is there such a thing as a "self-made" person? If there is, I haven't met them yet. In my case, too many to count had a hand in getting me where I am today. My family and my formative years at home prepared me with the right mindset to make the decision that changed my life.

I grew up in Youngstown, upstate New York. My parents divorced and remarried, so there were siblings and half-siblings and step-siblings, a dozen of us in all. First thing we'd hear in the morning was Mom's voice.

"Get up!" she'd call. "Eat and get out. Go blow the stink off you."

We didn't spend much time in the house, but we knew we had to be home when dinner was ready. We knew the consequences for being late: with such a big family, the food went quickly. If we were late, we got cereal.

During the school year, I did homework, had after-school activities, and hung out with friends. In the summer, I was never home—girlfriends, sports, and parties kept me busy. My stepfather owned racehorses, so we took family trips to Buffalo Raceway and

Batavia Downs. We also went across the border to see Niagara Falls from the Canadian side, where the gas and beer were cheaper.

I worked at my stepfather's bowling alley, played in neighborhood pickup games, and got into fights and neighborhood brawls. My brothers and I set the dogcatcher straight a time or two—we weren't going to let anyone take our family dog, even if he wandered all over town like we did. To balance that brash behavior, I'd get a double dose of church when I'd stay with my dad and step-mom, Diane, on the weekends.

Playing sports, especially team sports, was my thing. If it had a ball, puck, stick, club, or bat, and required hand/eye/foot coordination, running, or any kind of physical contact, I was there. I always looked forward to an organized traveling league, high school practice, or spontaneous neighborhood backyard game. The coaches and even the pickup team captains demanded perfection and hustle; it was always "do it again" or you sat the pine. I thrived on competition, and hated to sit out or lose a game. And the friends I made—well, they're for life.

No surprise, then, that I was attracted to the military lifestyle. I liked the discipline, its aggressive nature, and the team-oriented way of life. The greatest American team is the Marine Corps. It's the most physical, has a bias for fighting, and when the leadership of the country needs it done—or they aren't exactly sure what needs to be done—they send in the Marines. They know Marines always figure it out and carry the day.

So I signed up at 17, not long after Ronald Reagan was sworn in as president. Americans were no longer held hostage in Iran, and the "do you believe in miracles" U.S. hockey team had won gold at the Olympics. America's spirits were high, and so were mine.

*　　*　　*　　*　　*

Walk your post from flank to flank and take no shit from any rank.
A Marine on duty has no friends.
~ Unwritten 12th general order in the USMC

The general orders of a sentry were the first instructions my fellow recruits and I learned as poolees in the delayed entry program.[1] Our drill instructors hammered them into us on day one and demanded our resolute conduct when enforcing them. I learned that to be "on duty" doesn't just mean you're standing post or following orders. Marines are *always* on duty. That's just the way of the Jarhead. Our DIs taught us that a Marine is always standing tall, always hyper-vigilant, always faithful in doing the right thing, and always ready to assist under any circumstances. They taught us the right way to give and follow orders.

I also learned the value of discipline and structure. With our DIs warmly encouraging us, we were up and moving at zero dark thirty. The lights would come on and we'd hustle with intensity to get our heels on the yellow painted line that ran down both sides of the squad bay. We'd stand in our skivvies at the position of attention, staring across at another recruit.

---

[1] The delayed entry program uses the holding time a recruit has between when they make the commitment to become a Marine and when they actually ship to recruit training. During this time, recruiting stations motivate and prepare their poolees both mentally and physically for the rigors of recruit training.

"Rifles, move!" the DI would bark.

Every recruit would unlock their weapon, grab a full canteen of water, and get back on line.

"Count off!"

We'd count off to make sure everyone was present and had their rifle.

"Drink!"

We'd hydrate up and finish a full canteen.

"Secure 'em now. Do it!"

We'd lock our rifle back up.

"Port side, head calls now. Move it!"

When they hustled back, then it was starboard side's turn for head calls. We'd get dressed and make our racks while the DI was counting down.

"60, 59, 58, 10…"

He would skip a lot of numbers to motivate us. Within 10 minutes we'd be heading outside to form up, ready to march to chow. Twenty minutes later—yes, you can get 60-80 recruits from the squad bay to the chow hall, through the chow line, seated, breakfast eaten, and back to the squad bay in 20—we'd be doing morning cleanup. Exactly 60 minutes later we were double-timing to the PT field.

I left a couple things off that were also done in that hour: reveille (accountability), hygiene (head call and shave), racks made, canteens filled, chow, field day (entire squad bay swept and swabbed; head cleaned, including toilets, urinals, sinks, laundry room, and ladderwell; and outside police call), fire watch posted (guards assigned), sick call (if you were hurt or not feeling well), and day bag packed (after PT you move right to the next event, so the day bag has your recruit knowledge, camouflage uniform, boots, etc.).

Our entire day was like that: training event to training event until taps and hitting the rack. Repeat for three months.

One of my DIs inspired me and made me want to inspire others. He was a sniper. Most platoons are turned over to the primary marksmanship instructors during range week at WFTBN[2] who do all the teaching, but every night after snapping-in and shooting, our DI would spend more time with us. He would put us into the off-hand, sitting, kneeling, and prone positions.

"Breathe!" he would shout. We'd inhale deeply.

"Relax!" We'd exhale and make sure there was no bone on bone.

"Aim!" We'd focus on the front sight tip, good stock weld, good eye relief.

"Stop." Natural pause, easy now. "Squeeeeeeze."

Click. The hammers dropped.

"What was the last thing you saw when you squeezed?" he'd ask. "Don't bullshit me, I'll know tomorrow."

If any recruit didn't see a clear front sight tip and a hazy dark silhouette in the background, our DI would immediately come over and stand, sit, kneel, or get on his face with him.

"Do it again," he'd say. Every morning at lights on and every evening at lights out, we recited the Rifleman's Creed.

---

[2] Weapons and field training battalions

I rehearsed his lessons, the basic marksmanship fundamentals, over and over in my head. I knew then and there I wanted to be like him. Not just a sniper, not just a rifleman with an advanced skill set, but someone who demonstrated and took to heart the enduring trust qualities—someone who was trustworthy, committed, dependable, moral, ethical, and knew how to work as a team.

As my fellow recruits and I developed through boot camp and the operational forces, sergeants and corporals taught us not just to be unwavering in executing our orders, but also to treat all with fairness, dignity, compassion, and respect.

*   *   *   *   *

Three years later, still on my first enlistment, I got married. Within the next two years, my incredible wife, Susan, and I were blessed with two baby boys, Michael and Jonathan. Since then, we've moved 16 times—which doesn't count the four times I moved alone to a permanent change of station so as not to take our boys out of school. For 34 years, we've been together through multiple overseas and combat deployments, field exercises, combined arms exercises, pre-deployment training packages, temporary additional duty orders, duty days, and lost weekends.

I knew that being inconvenienced and uncomfortable would come with being a Marine. My family knew it too, and they were up to the challenge. We are stronger for it. I made as many graduations, baseball and soccer games, and wrestling tournaments I could for my boys, and I was home for most holidays, birthdays, and anniversaries. The ones I missed were because I was deployed, in the field, training, or otherwise assigned elsewhere, doing my job. My family will always take priority when they absolutely need me. If things at home were tracking in a forward direction and Susan was taking lead and all was well, then the Marine Corps had my attention.

Susan almost singlehandedly raised our children, and she led our home with integrity, order, and a natural gift for leadership. She made sure our family was fit emotionally, socially, financially, and spiritually. She was the shoulder I could lean on, the voice of reason I could listen to, and the foot in the rear I needed.

She is still all of those things today, and Michael, Jonathan, and I praise her for all she has sacrificed for our family. Truly, she's the epitome of the virtuous woman described in Proverbs: "She openeth her mouth with wisdom; and in her tongue is the law of kindness. [...] Her children arise up, and call her blessed; her husband also, and he praiseth her."[3]

She allowed me to be a Marine, as she promised. And I was always there when she needed me, as I promised. The single most important decision a person will make in their lifetime is the person they choose to be their mate. When I talk to young Marines, this is what I tell them: "The most important thing you'll ever do in your life is be a Marine. The most important thing you'll ever *have* in your life is your family." ■

---

[3] Proverbs 31:26, 28 (KJV)

Lesson One: Checklists and Habits

Good checklists [...] are precise. They are efficient, to the point, and easy to use even in the most difficult situations. They do not try to spell out everything—a checklist cannot fly a plane. [...] They not only offer the possibility of verification but also instill a kind of discipline of higher performance.
~ Atul Gawande, M.D.

I was a 19-year-old corporal with 1st Battalion, 4th Marines, in Twentynine Palms, California, when I received orders to the Army Rangers School. I was on a sniper team, a member of the Surveillance and Target Acquisition Platoon. I knew Ranger School would test me like never before in my young career, and I expected to learn to Soldier as a leader while enduring the great mental and psychological stresses and physical fatigue of combat.

What I didn't know was how checklists would change the way I did things for the rest of my life.

One morning, the Ranger instructor (RI) in charge walked into the patrol base and called for the patrol leader and Ranger Barrett. At this point, I was serving as just another member of the patrol, not in a graded billet.

"Patrol leader, you're relieved," the RI said. "Conduct an Assumption of Command brief."

We quickly pulled out our field notebooks, where we'd handwritten checklist after checklist. We flipped to the Assumption of Command fail-safe checklist, followed the procedures, and everything went perfectly. It's hard to screw something up when you have the instructions right in front of you.

Ten minutes later, after I walked the perimeter and checked security, the RI in charge handed me an intelligence summary report and a frag order—a new situation and mission.

"You have two hours to complete the planning phase of this mission," the RI instructed. "Tick tock!"

Suddenly, I was in charge and under the microscope. Everything our patrol did or failed to do was on me. I would be graded on my ability to plan the entire mission, including where, when, and how to occupy the next patrol base.

I mentally followed my checklist. I prepared and issued orders and conducted rehearsals. We did everything you can imagine to ensure the patrol's success: order of movement, actions on enemy contact, actions on the objective, casualty evacuation, crossing a large open danger area, crossing a water obstacle, and so on.

After rehearsals, we conducted a detailed head-to-toe inspection of each member of the patrol. We also checked common gear and special equipment for certain members and fire teams. The level of detail in this phase alone is staggering: you must ensure everything is serviceable and secured to gear, there's nothing shiny or rattling, that all faces have camouflage paint, canteens are full, weapons are clean and working properly, communications gear is operational, batteries are fresh, communication positive with higher HQs and fire support assets, and much more. The checklists for movement to the objective and actions at the objective phases are even more detailed and demanding.

Battle, and preparing for battle, requires detailed planning; a good checklist leaves nothing to chance. The longest patrol order I ever wrote was a sniper mission during the Gulf War. It was 117 pages long.

Over the years, I became a checklist master. While in uniform, I used a Plan of Action and Milestones (POA&M) checklist for events planning:

- *What* (issue/item/event)
- *When* it needs to be coordinated and/or accomplished (no-later-than date)
- *Who* is responsible
- *Status* column for quick reference—green, yellow, red (too much red means I have not done much and am probably behind)
- *Remarks* column for forget-me-nots

I still use that checklist today. I also still use a problem-analysis decision matrix (PADM) with core functional leadership actions to solve a potential issue or unforeseeable setback (crisis/predicament/mess/dilemma).

Even retired, the start of my day has a checklist—a basic daily routine (BDR). I get up, make the rack, make a head call, read (the Bible, a non-fiction book, or a thriller), take my multi-vitamin, drink a 32-ounce bottle of water, eat a protein bar, work out, clean up, eat breakfast, catch up on the news and read some more, think about what I just read, have a cup of coffee as I do, and then whatever I have (or my wife has planned for me) to do. All this occurs before 10:00 a.m. Good habits die hard. Checklists helped me develop those habits—and good habits forge strong leaders and disciplined warfighters.

As Sergeant Major of the Marine Corps, I traveled the country and the world to visit our Marines. I never visited a unit with the intent of finding something wrong; I always went searching for Marines doing great things. But I was constantly making assessments and noticing habits and attention to detail as I moved about.

I was always able to sum up the leadership in a unit pretty quickly. I would observe how the Marines carried themselves: their haircuts, how they wore the uniform, the way they saluted, the pride and upkeep of the barracks, the command deck. Were the bulletin boards current, or were there outdated orders, training schedules, and duty rosters? Were the motor pool, armory, chow hall, and common areas neat and organized?

Every time I'd see trash on the ground or something out of place—like an MRE lying haphazardly in the sand at a forward operating base—I immediately knew the unit was sloppy, reckless, and unfocused.

"You know what that tells me?" I would say to them. It wasn't a question. "It tells me either you're fucking lazy or you don't care. Think what you're doing when you leave the wire.[4] Where's your effort and focus?"

I would stick it right in their hearts to make my point clear. Carelessness gets people killed; you have to establish good habits, attention to detail, and order from the beginning.

I would ask to see the unit training report, because it showed physical fitness and training standards readiness. I would review the legal report, for it exposed negative trends. I would read the Binnacle report, noting how many sickbay commandos there were and

---

4 "Outside the wire" is military jargon for going beyond the relatively safe confines of a base camp or support installation.

MICHEAL P. BARRETT | BREACHING THE SUMMIT

whether that number increased on days of battalion/squadron long runs or force marches. Were there contributing factors that revealed motivation, morale, and esprit de corps issues? I would review the career-planning report, which showed reenlistment information. Was the unit making retention mission? If not, why not?

All these reports, along with my first-hand look at how the unit walked, talked, and dressed, revealed the unit's morale, proficiency, esprit de corps, discipline, and motivation. If I found trends, good or bad, I would discuss the details with the leadership to better understand what they were doing or not doing. I'd always do a hot wash (a "lessons learned" debrief) with that unit sergeant major to share my notes of the things I liked and the things they could do better. Then I would follow up: I would return months later with the same notes in my pocket, and I'd make sure they had fixed the problems.

"The nature of men is always the same," Confucius said. "It is their habits that separate them." My fellow Marines and I rigorously adhered to our checklists because sticking to them developed discipline and good habits. We didn't follow them just because we wanted to be effective leaders; we were motivated to do so because we knew that these checklists would, in many instances, save lives. ■

Lesson Two: Character

Leaders are made, they are not born. They are made by hard effort, which
is the price all of us must pay to achieve any goal that is worthwhile.
~ Vince Lombardi

Everything I need to know to survive in life I learned before I was 21. By then, I had been in the Marine Corps three and a half years. Everything I have learned since has been nothing more than closing a knowledge gap of understanding. I watched, listened, read a lot, got the wind knocked out of me, scribbled notes, and repeated over and over in my head so as not to forget what someone had just taught me.

Many times the lesson came simply by observing a leader's character. I had to become a great follower before I could lead. I had to become a human *being* before I could become a human *doing*.

The 14 leadership traits of the Marine Corps are all connected under the umbrella of character: bearing, courage, dependability, decisiveness, endurance, enthusiasm, initiative, integrity, justice, judgment, knowledge, loyalty, tact, and unselfishness. I add two more traits to that list: humility, because it keeps us on track and makes it possible for us to see past our own impulses; and cheerfulness in the face of adversity, one of the unique characteristics of the British Royal Marine Commandos, some of the world's most proficient warfighters. They are happy warriors.

Throughout my career in the Marine Corps, particularly on my tours in Afghanistan and Iraq, I have seen countless examples of each of these character traits. Several acts of valor and sacrifice stand out in my mind.

In June 2007, Lance Cpl. Ian Dollard was on patrol with his team in Saqlawiyah, Iraq, when they were ambushed. His platoon commander, 1st Lt. Paul Brisker, was hit by enemy fire. Lance Corporal Dollard threw himself on top of his commander, shielding him from further fire and administering first aid. Dollard took two rounds to his back armor but dragged his lieutenant more than 25 yards to safety while under fire. Dollard was struck again, this time in the leg. He ignored his injury, making sure the platoon was supported on their way out and back to an operating base. Only then did he tell his staff sergeant, "Oh, by the way, I got shot too." First Lieutenant Brisker made a full recovery, and Lance Corporal Dollard received the Silver Star for his selfless actions, courage, and quick thinking.

On July 18, 2010, Cpl. Joe L. Wrightsman led a partnered Afghan National Police patrol conducting a recon on the far side of the Helmand River, which was running high and swift. Suddenly, one of the ANP was caught up in the current and swept away. Without hesitation, Wrightsman jumped in after him—even though he couldn't swim and both men were wearing 100-plus pounds of combat gear. They were no match for the strong current, and both men drowned. Why would he risk his life for a man he didn't know who spoke a language he didn't understand? Because selflessness and bravery were part of his character, and such a fearless action was second nature. Wrightsman was posthumously promoted to sergeant for his outstanding courage and sacrifice.

Doing the right thing is the compass that guides Marines and shapes their characters. It's the cornerstone of our duties and responsibilities. Thomas Edison said, "What you are will show in what you do."

The higher leaders rise, the greater the responsibilities and the greater the sacrifices they will be asked to make. Leaders must always be taking and making assessments and raising the standard.

I size people up by asking the following questions:[5]

- Are you **competent**? Do you finish everything you start? Do you demand excellence from yourself and those in your charge?
- Are you **committed**? Is everything you do for the team's success? Are you focused on the big picture? Do you have a whatever-it-takes attitude? Have you bought into the spirit of the unit? Are you self-motivated and productive? Do you have passion and enthusiasm? Do you arrive early and leave late?
- Are you **dependable**? Can you be depended on every time—not just when it's convenient, but every single time? Do you keep your word? There is no greater compliment than to be told, "I can always count on you."
- Are you **ethical**? Is your integrity unquestioned? Do you do what you say you'll do? Do you give of your time and watch out for your seniors, subordinates, and peers alike?
- Are you **team-oriented**? Do all your actions bring the team together?

I've shared this list with thousands of Marines. Those are the qualities I look for in a leader.

Those we lead will follow our example. The way we walk, the way we talk, the way we dress—it all matters. Our subordinates hear and see everything we do and say. They are, after all, a product of our leadership. If we do it, if we say it, it must be OK. And they will repeat our actions times ten. Before you do or say anything, ask yourself: Is what I'm about to do or say immoral, unethical, unprofessional, unsafe, unjust, or illegal? If you have to pause for even a nanosecond, don't do it.

Scottish author Samuel Smiles wrote, "The great high road of human welfare lies along the old highway of steadfast well-doing."

*   *   *   *   *

Gen. James F. Amos, 35th commandant of the Marine Corps, is a walking example of every trait that characterizes an effective leader. For almost four years, I had the privilege of serving as his senior enlisted advisor. He's a wonderful man and exceptional leader who loves his Marines.

He showed loyalty, humility, and tact in the way he communicated with me—especially the handful of times he corrected me. Always a gracious gentleman, General Amos would wait until we were alone.

---

[5] I give credit to the gifted author of many great leadership books, John Maxwell, for his philosophy on enduring trust qualities and indicators of effective leadership.

"We could have done that a lot better," he would say. He always used *we*. "Let's talk this through. How are we going to handle this the next time?"

And then we'd go out there and he and I would knock it out of the park. I cherished our interaction.

We were fighting on two fronts during those years. The Budget Control Act, continuing resolutions, and sequestration were now the norm. We had $2.6 billion less in our budget. We were forward deployed across the entire globe and told to draw down while we were still in direct action with the enemy.

General Amos showed decisiveness and courage in how he handled those most difficult times in our Marine Corps history. Under those budgetary restrictions, balancing the five pillars of institutional readiness was an enormous challenge: motivating and retaining high-quality people; unit readiness; equipment modernization; infrastructure sustainment; and the commandant's responsibility to provide capability and capacity to Marine Corps commanders around the world.

General Amos is one of the smartest people I've ever met. He is cerebral and has extraordinarily vast experience. He also has what the Germans call *fingerspitzengefühl*: an uncanny, intuitive instinct. He was the right leader at the right time. He saw where we needed to make course and speed corrections, analyze information, and get the right data so we could effectively lead the Corps during those turbulent times.

"Let's do this right," he would say. He was patient and measured.

We went through social and total-force evolution together: review of the restriction on women in service, implementation of the Don't Ask Don't Tell repeal, unisex uniforms. We were juggling 79 glass chandeliers, and you couldn't let one of them drop.

On top of all that, some of our young Marines sometimes made poor life choices, and the challenge of keeping everyone "in the fight" (deployable) weighed heavily on both of us. General Amos's approach to quality of life in the Corps was "eyeball to eyeball" engagement. He wanted his arms wrapped around the problem, and the only way to do that was by being directly involved at the front-line level.

General Amos made sure that every Marine knew the state of the world and that it was not getting any nicer. He and I laid out who we are as a Corps, what we do for our nation, and what our responsibilities are as a war-fighting institution. We traveled the world to talk to every single one of our Marines—more than 200,000.

The commandant's positive attitude and enthusiasm were contagious, and his leadership had a significant impact on our Corps. In essence, General Amos let the Marines know that he fully understood the reality of the challenges, but they weren't going to tear us down.

Yep, we're fighting on two fronts. Yep, our budget has been cut. We're going to have to do more with less, but we are not going to do it less well.

General Amos made it clear that Marines were not going to take a step back.

"I need you to bring your 'A' game," the commandant would say to our Marines.

General Amos was laser-focused on keeping every one of our Marines in the fight, as we could not afford to lose a single one to poor life choices.

When I spoke, I would phrase things a little differently.

"Every Marine deserves to be in a good unit, and led morally, ethically, and professionally," I'd say. "We live hard, train hard, and fight hard. We don't have time for this backdoor alley-cat bullshit, drug use, alcohol incidents, domestic violence, criminal mischief, sexual misconduct, hazing, reckless jackass behavior, or self-destructive thoughts. We need every one of you in the fight! If things in your life are getting away from you, tell someone! Tell the commandant! Tell me! We'll get the support you need."

General Amos was always taking stock of how we were doing as an institution. I told him our EPME (enlisted professional military education) needed an upgrade. It needed to evolve.

"OK, let's re-write it," he said. He gave me the latitude to bring our EPME up to date and tie it to promotion requirements.

Together, we worked 29/9. He had 15 years on me, but he maintained an enviable zeal and vigor to do his job. I didn't need to look very far when I was feeling run down; he was the example to follow. He's every single one of those leadership traits we all try to emulate.

We traveled almost every week of our time together. On the plane, he and I would sit next to each other, answering emails on our laptops while we talked about our plans once we landed. But when we needed a break, the commandant didn't recline back and take a nap—he would reach into his bag, grab a book, and start reading. He was constantly learning more and keeping his brain engaged. The light switch was always on. He was 60+ years old, at the top of his career in a high-intensity environment, but he never slowed down. Of course, if the Secretary of Defense or Secretary of the Navy were taking a long weekend, then we would take advantage of that rare opportunity too.

"Let's re-arm, re-fit, and re-deploy in about 96 hours," he would tell me.

Despite his brutal schedule, General Amos and his wife, Bonnie, frequently opened up their home for Marines and their families and friends, hosting dinners, receptions, and meetings. Even though Corps business was the priority, guests at the Amos home always felt like they were the most important people there. The poet Maya Angelou captured what was so unique about that: "I've learned that people will forget what you said, people will forget what you did, but people will never forget how you made them feel."

He had an institution to lead, and nothing was a greater priority to him than the Marine family. When the history books are finalized, you'll find that General Amos shepherded the Corps through one of the most tumultuous times in our 242-year history. It fell upon his shoulders, and how he carried himself was exemplary. The Marine Corps leadership traits and principles are tattooed on him. I'm a better man because of him. ■

Lesson Three: Cultivate Your People

If you give a man a fish, you feed him for a day.
If you teach a man how to fish, you feed him for a lifetime.
~ Proverb

Growing up, I spent a couple of summers working on farms and vineyards. I used to shovel manure, so when it comes to poop, I am some kind of expert. What happens when you spread manure, cultivate the earth, and water the ground? Things grow. But take that same pile of manure and stick it in the corner of a barn, and now it's just a big pile of stinky poop sitting around doing nothing. My point is, leaders have to spread themselves to be effective, and they have to cultivate and water the people they work with in order to grow future leaders.

"Be like manure" is my unofficial credo. You do that by first putting into practice the Marine Corps' leadership principles, and then implementing these suggestions:

1. Set them up for success
Thinking outside the box is a valuable skill, but people have to know what's in the box before they can think outside it. Help your people master their occupational specialties. Set them up for success by ensuring they can assume the duties and responsibilities of the next grade or next-higher billet.

As a first sergeant and subordinate sergeant major, I made sure the Marines in my unit always attended the necessary courses, schools, and academies. As Sergeant Major of the Marine Corps, I sent every member of my team to the required school they needed to do their next job. Yes, I sometimes had a shortage or two (or three) on the staff, but I was confident my team was cross-training with each other, and everyone could do everyone else's job in the event of a gap. The important thing was that my Marines got the schooling they needed so they could grow.

Our leadership academies and military occupational specialty and advanced proficiency schools are top tier—the finest instructors teach everything we need to know and be able to do. As your people get promoted and elevated in billet, get them to the next school ASAP!

Another way to set them up for success is to make them feel like they're part of an outstanding team. I once spent time with a company that won every Marine of the Quarter and NCO of the Quarter awards for a year. Additionally, the company commander was a Leftwich Award recipient. I asked a corporal if he was one of the award recipients, and his response blew me away.

"No," he said. "I wasn't one of the most valuable players, but I was still on the championship team."

Making each member of your team feel like they're part of the success is how you build morale and inspire people.

2. Give them responsibilities and accountability
I had to grow my bench early, because I knew I wouldn't always be around—I would need

to travel with the commandant. I defined the team's roles, clarified expectations, and gave them responsibilities that made them stretch. Accountability, of course, was key: I gave them the authority commensurate with the task, and they owned the consequences, both good and bad.

I could go on nine-day road trips with General Amos with no worries, because I knew that my team could—and would—handle anything that came across my desk while I was gone. I was confident about that because I had cross-trained with them, and they knew what a good job looked like. Unless they asked me, "How do I do this?," I left them alone to their own skills, capabilities, guts, and know-how. They never once disappointed me!

Taking on new responsibilities is difficult for everyone at first. Nonetheless, your team will be stronger, faster, and smarter when everyone has been given the opportunity to lead or has been placed in a position outside their comfort zone. Properly trained Marines are more likely to step up and carry out important tasks if they have been given the opportunity to be resourceful and to develop and expand their leadership skills.

But don't tell people how to do things—just tell them what you need done. Don't meddle with them while they're doing it. Don't manage your Marines—lead and inspire them! If you want to manage something, go down to Supply and manage the inventory, or go to the armory and count the PEB or the motor pool and see what it's like to change a tire on a seven-ton truck in Twentynine Palms in the sun on a 100-degree day.

In 1996, I was the company gunnery sergeant and training chief at Camp David. One day, I was teaching tactics and procedures to one of the platoons. I was systematically running through the procedures when the CO and the first sergeant showed up to observe the training. Everything was going the way it was supposed to, so I told the platoon to keep going and I walked over to shoot the breeze with the CO and first sergeant.

1st Sgt. Mark Gordon was an infantry guy, and like any good infantryman worth his weight, he wanted to give me some advice on the training.

"Here's the method I think you should be using," he said, and started giving me some instructions.

I nodded respectfully and told him thank you.

"This will get to the end state," I assured him.

"But my technique and method works," he insisted. "Here's why."

I let him explain. I am a man of equanimity, but this tested me.

"Either you're doing the training or I'm doing the training," I finally said. First Sergeant Gordon understood and put up his hands in the universal gesture for "I surrender."

"You're right," he said. "Fair enough."

They left the training area, but later that day, the first sergeant approached me.

"You're doing a great job," he said. "You were right to say what you did. As the CO and I were walking back earlier, he said, 'If you need to inject yourself in the training, go ahead.'"

First Sergeant Gordon shook his head.

"I told him, 'Sir, Gunny Barrett is doing exactly what I would have done. When I was a young company gunnery sergeant, my first sergeant trusted me—he knew I was going to get it done.' There's a hundred ways to skin a cat, and what matters is the end result."

In other words, we both had our own way of training, and both ways were right.

"When has Gunny ever let you down?" First Sergeant Gordon asked the CO.

He taught me about trust and accountability that day, and he taught our commanding officer, too.

3. Don't solve their problems

Good leaders don't babysit, coddle, or handhold. They are demanding and firm but fair, and they teach you to be responsible for your own actions. For the same reason, I refused to be a problem-solver for my Marines.

This is how I would handle someone with a problem: I'd tell them to sit down and tell me what was going on. After they explained, I'd ask, "So what do you think you should do?" They would give me their ideas, which were usually well thought out.

"I think that sounds worth doing," I would say. "Let me know how that works out."

All they needed was a sounding board—I never solved their problems.

"Listen to me," I'd say, looking a Marine in the eyes. "I trust you. You had the answer all along."

However, from time to time I did need to suggest a plan B. I remember this young lance corporal was habitually showing up late for formation. One day, a charge sheet for UA appeared on my desk. I called the platoon sergeant in and asked why the young Marine was always late.

"His car broke," the sergeant responded. "I told him to get it fixed and stop being late."

That wasn't the entire story, though. After peeling back the onion, I learned that the problem wasn't negligence or failure to follow simple instructions—it was money. The lance corporal and his wife and toddler were living paycheck to paycheck. They lived off base, and his wife couldn't find a job that covered the cost for childcare and other necessities. The lance corporal relied on friends who lived close by for rides, but as they were from different units, they reported at different times for unit formations.

I asked the platoon sergeant what he thought we should do.

"I'll loan him the money and he can pay me back," he said.

That was generous, but I wasn't convinced it was the right solution. I mentioned the Navy-Marine Corps Relief Society and the Armed Services YMCA on base, since they could help a struggling family. Then a light bulb clicked on.

"Does he know how to fix his car?" I asked the sergeant. "This is why we have an auto hobby shop!"

That same day, the young lance corporal was escorted to the NMCRS for financial assistance and signed up for a financial fitness class. That weekend, his platoon sergeant helped him get his car to the auto hobby shop, and the resident gearheads worked with him to get his car running. All it cost was money for the parts. *Teach a man to fish, and you feed him for a lifetime.*

The young Marine eventually got promoted to corporal. On most Saturdays, you could find him at the auto hobby shop, helping other Marines in need.

4. Know your people

Take the time to get to know your people. Learn what's important to them. They will remember you were there and they'll recognize that you made great efforts to share your time.

I liked to stay in the barracks when I traveled. You want some cold hard facts? Spend a night with a couple of lance corporals. Two lance corporals turn into a dozen or more in less than 10 minutes. Hang out in the beer garden while they're playing beer pong, or just take 15 to 20 Marines to the chow hall. I always had a list of questions to ask, and I never had a shortage of raised hands and input. I believe I had the pulse of the unit.

When you're talking with one of your Marines, connect with your eyeballs and earholes: look them in the eyes and listen to what they're saying. Pay attention to their tone and disposition, too.

In boot camp, my DI handled every recruit with firmness, fairness, dignity, compassion, and respect, as it was due. In the seven cycles that I pushed (trained) recruits through the entry-level pipeline, I tried to do the same. I treated them firmly—my mission was making Marines, after all—but I watched for signs when they needed some compassion. I could tell when something was wrong; maybe a recruit got a letter from home saying his parents were getting a divorce, or maybe his girlfriend broke up with him. A young guy away from home for the first time, going through stressful recruit training, needs a surrogate father. When I could tell something was amiss with one of my men, I would tell him to come to my duty hut.

"Help me to better understand what's going on, recruit," I'd say. At that point, they'd often let their emotions get the better of them. I'd listen to what was going on in their lives, then I'd let them stay for a while until they were composed.

"Go use my private head and come out when you're ready," I'd say. I would never throw an emotional recruit back on the quarterdeck.

The DIs I had in boot camp were harder than woodpecker lips, but they took care of us. My recruits say that I was harder than beaver dentures, but I treated them like sons.

5. Take care of them

The Survival, Evasion, Resistance to Interrogation, and Escape School (SERE) equipped me for sacrifice—not just to give my life for others, but how to forego creature comforts. I learned to eat or sleep only *after* I had taken care of those I was responsible for.

Late one night in 2006, one of my platoons came in off a three-week operation in Iraq. The Marines were tired, dirty, and hungry, and they headed straight over to the chow hall at Camp Fallujah to get some warm food in their bellies. They didn't even stop to shower. A corporal at the chow hall told them the place had closed down ten minutes before.

I was not going to let anyone turn away one of my hungry Marines who had been fighting the enemy. I swooped in there and raised holy hell.

"Here's the deal," I spit out. "You are going to feed my boys right now. In fact, you're going to serve them yourself."

The corporal fixed their plates, and my platoon got fed that night.

Part of taking care of your Marines is also helping them make the right decisions. Use liberty briefs, for example, to remind them that nothing good happens after midnight. Be specific. Remind them of the actions that will destroy everything they've worked hard to earn and will devastate their family and friends: drugs, alcohol, domestic violence, criminal mischief, sexual misconduct, force-preservation deficiencies ("jackassery," I call it), hazing, suicide indicators, POVs, and lack of motorcycle defensive-driving vigilance.

I would regularly ask my Marines: "What are you doing this weekend? Where are you going? Who are you going with? What's your link-up plan if you get separated? Do you have my number? Call me if you can't get ahold of your battle buddy!"

I once told a group of Marines, before cutting them loose for liberty, that if anyone wanted a hangover, I could just stick a dirty sock in their mouth and punch them in the stomach—it would have the same effect.

Always be working toward ensuring the best quality of life for those in your sphere of influence. Quality of life, to me, means bringing them home safely—whether it's from a deployment or a long liberty weekend.

6. Be a leader-servant

I pay attention to how senior enlisted leaders describe the number of Marines in their commands. If they say, "I am responsible for #" or "I have # in my charge," I am quick to remind them that they *serve* every single one of those Marines. It's not the other way around. We serve, they support, we win.

Don't forget to look at things through the eyes of a private, and never forget who's looking back at you. The people who wear the cloth of this nation chose this life. They could have gone and done anything they wanted. Always remember that it's a privilege and an honor to serve them as well as lead them.

Your team must see you in the trenches getting dirty and sharing in the hard day-to-day efforts, not just making savvy leadership decisions from behind a desk or handing out an award for Marine or NCO of the month.

7. Enlarge them at promotions

Balian of Ibelin was a crusader noble of the Kingdom of Jerusalem in the 12th century. Having inherited the responsibility of defending Jerusalem during Sultan Saladin's attempt to take the city, Balian began knighting commoners, peasants, or anyone who could wield a broadsword. Patriarch Heraclius denounced what Balian was doing because the people were not worthy of knighthood, he said. They were not royalty and they were untrained. However, Balian defended his decision by explaining that knighting a man made him a better fighter. He and his small army defended the city and saved the lives of all those within its walls.[6]

Balian empowered these men and showed them great respect for what they were about to do. Because he made them feel like knights, they fought like knights.

I'm not suggesting we mass-promote anyone willing to fight or that we dismiss the service and grade requirements, but *how* you promote and conduct a promotion ceremony is crucial to growing strong leaders. I was taught early in my career that you share the accomplishments of the Marine being recognized with everyone in attendance. The senior enlisted leader, usually a first sergeant or sergeant major, explains to the newly promoted person his or her responsibilities and what the appointment to the next grade means. What

---

6 In 2005, Ridley Scott directed the 20th Century Fox film "Kingdom of Heaven," a fictionalized portrayal of the life of Balian of Ibelin.

was most meaningful to me was being told to "always wear it well, demand excellence, and enforce the promotion warrant no matter where you are." If we enlarge our NCOs at promotion ceremonies, they will be revered and treated like NCOs; they will fight like knights.

Know what fills a Marine's motivation tank, and keep it full. Rewards come in different shapes and sizes: some want time off, some want recognition and praise, and some need public accolades or plaques. My favorites are the ones who just want more responsibilities.

8. Treat your people with dignity and respect
- The golden rule always trumps the egotistical, loudmouthed jackass.
- Pull people—don't push them!
- Give your personal attention and time to those who have made honest mistakes but are working hard and doing their best. One day when I was sergeant major of the First Marine Division, I was walking to the MCX when I saw a staff sergeant yell at a young Marine who had his hand in his pocket (a no-no while in uniform). Although it was the staff sergeant's responsibility to correct a violation of the uniform regulations, what boiled my blood was that he yelled at a Marine who was outside enjoying lunch with his wife and little boy.

I walked over and introduced myself to the young family. I thanked the Marine for his willingness to serve and his wife for her sacrifices, and let her know how much I cherished their endurance. I asked the staff sergeant to walk with me. Within just ten yards or so, we encountered a couple more Marines keeping their hands warm.

"Watch," I told him. I approached the group with my hand and arm extended, and immediately their hands came out of their pockets and we all shook hands. We engaged in some ooh-rah talk and Marine bonding. Without causing friction, I had subtly reminded them where their hands should be while in uniform. While we were talking, we discovered that one of the Marines was from the same hometown as the staff sergeant. They talked, laughed, and exchanged numbers.

"Spilled milk," I call it. What happens when a child knocks over their milk? You smile, tell them it's going to be OK, wipe it up, and move on. No need to get loud, verbose, or testy.

- Respect your people's schedules: start meetings on time and make sure they're well planned. I usually got to the office at 0530 every day during my tenure as SMMC, but my staff beat me there every day. On top of that, more days than not they were still there in the evening after I left. I was busy, but they were busier making sure I was supported. When I needed to talk with my team—Master Gunny Cypress, First Sergeant Lillie, and Gunny Nuntavong, to name just a few—we had "stand up, in, out, done" meetings. The key is to be prepared: stand up (no need to sit; you won't be here that long), check in (discuss what's needed or happening, when it needs to happen, and who is running lead), check out (thanks for your efforts; let me know if you need my assistance), and done (now get out—you're wasting time). I was working 29 hours a day, and my team was also. They have families and lives, too. Don't waste their time; they're just as engaged, busy, and important as you are. ■

Lesson Four: Optimal Combat Readiness

*One secret of success in life is for a man to be ready for
his opportunity when it comes.*
~ Benjamin Disraeli

Our short life is a contact sport. You're going to get the wind knocked out of you, you're going to taste and see your own blood, you're going to get screwed over, you're going to have your wallet grow legs and run away, you're going to get weekend duty, and someone you serve with will be taken from you in the most horrific manner. Be ready for it so when it does happen, you won't be shocked and you'll react appropriately. Prepare for life's unexpected moments. Live in the *ready*, not in the *hope* of what will be.

Optimal combat readiness is a concept that can apply to anyone—it's not just for the military. If you're a civilian, optimal readiness implies being prepared to take a job promotion on short notice. If you're a Marine, it means you're ready to deploy or engage in combat with little warning.

"Rely not on the likelihood of the enemy's not coming, but on our own readiness to receive him," the Chinese general Sun Tzu wrote in *The Art of War*.[7]

The measuring stick for gauging your level of readiness is comprehensive fitness: Are you physically fit? Mentally, emotionally, personally, and morally fit? Because a leader teaches by example, you need to make sure that you can easily check off the boxes for each level of fitness.

Physical Fitness

Anyone can score a high first class or perfect score on a personal fitness test—just get off the computer, walk away from behind a desk, put down the junk food, and train. Being physically fit to me is being "combat fit"—I won't accept anything less.

Are you able to put on 100-plus pounds of body armor and equipment and move 10 to 15 kilometers? Being on patrol requires you to be hyper-vigilant and requires two times the effort and energy of just moving in 100-degree temperatures—and you must still have the fortitude, endurance, and strength to fight the enemy by fire and maneuver. That's the standard I always trained to. Even today, retired and in my 50s, I still do medieval physical training.

Mental Fitness

Are you mentally agile, able to do multiple tasks in a chaotic environment, making critical decisions at critical times?

I remember the quick-thinking actions of Staff Sgt. Khris DeCapua, whose squad

---

[7] *The Art of War* is an ancient Chinese military treatise dating from the fifth century B.C. It is considered a definitive work on military strategy and tactics. John Murray translated the treatise into English in 1908; the first British edition was entitled *The Book of War: The Military Classic of the Far East [The Art of War]*.

was ambushed in Zaidon, Iraq, as they crossed a large, open danger area. One Marine was immediately wounded. With the potential for chaos all around him, DeCapua immediately started barking out orders: direction, description, range, fire assignments, fire control. He started maneuvering his reinforced rifle squad and got his radio operator on the hook with the rear sending reports on casualty evacuation and troops in contact.

They were uncovered, in the open, a Marine down receiving aid, controlling the fire and maneuver, sending in critical reports, and maintaining geometry of fire[8] to prevent collateral damage to innocent farmers in the area.

DeCapua directed a two-man 60mm mortar team, Lance Corporals Peffer and Guzman, to fire on the building where the enemy machine gun fire originated from. Exposed to enemy fire in the open, Peffer and Guzman flawlessly delivered timely and accurate suppressive fires from their 60mm mortar in a handheld position, disrupting and displacing the enemy and forcing them to abandon fortified positions as the maneuver element assaulted forward. DeCapua, at the ripe old age of 23, led and carried the day.

Are you mentally tough enough to endure sleep, food and water deprivations, sore shoulders, blisters, and extreme temperatures? Being mentally tough induces physical toughness, willpower, determination, mental poise, good health, and self-confidence. It means possessing a martial spirit, understanding that you will see blood, maybe your own; you will see and hear broken bones and screams, missing limbs, and possibly even death. And through it all, regardless of the situation or being outnumbered, you are thinking, "I am going to win!"

Emotional Fitness

Without emotional resilience, you cannot focus on a mission in the midst of a catastrophe.

On December 1, 2005, the 2nd Battalion, 7th Marines suffered a gut- wrenching tragedy when 10 Marines in Company F were killed by an IED outside Fallujah. Eleven more Marines from the same platoon were wounded.

The next 72 to 96 hours were enormously demanding for us. We were all grieving, but Company F still had a mission to complete. We needed the remaining 170+ Marines and Navy Corpsmen in the fight—not just their bodies, but also their cognitive abilities. We needed everyone alert, in control, focused, and agile. There would be a time for grieving and memorializing our brothers, but this was not the time or place. During those crucial hours after the explosion, we needed to gather intelligence and confront the enemy.

We did all the things you should and can do following something that horrific. All five of our first sergeants—our Tiger Team: Scott Hawes, Dan Fliegel, Joel Collins, Wayne Anderson, and Bill Sweet—ensured that all administrative requirements were accomplished expeditiously. Battalion commander Lt. Col. Joe L'Etoile and I spent the next three days preparing for a high-value target takedown. Of course, we still needed to make time to eat, rest, patrol, and even do some forward operating base PT with the Marines of Company F (including an intense "combat" dodge ball tournament, which means physical contact was not only anticipated but required).

---

[8] "Geometry" of fire is a military term that refers to the physical offset between the orientations and positions of friendly units on a battlefield. This offset must be properly maintained at all times in order to mitigate the potential for fratricide.

The reinforced squad-sized element remaining from that battle-hardened platoon would be in the assault, support, and cordon elements of the operation. In the end, we were successful: HVT[9] captured.

The weeks that followed set the tone for the entire battalion. We memorialized our brothers, told stories, showed pictures, shared non-alcoholic beer, toasted, laughed until it hurt, shed a tear, and took it to the enemy. They performed exactly the way you'd expect your Marines in combat to perform: they were steely-eyed, in-your-face, smash-mouth warriors. Every sergeant, corporal, lance corporal, and private first class stepped it up and were baptized in combat under the worst of circumstances. These Marines were mentally and emotionally resilient.

As the war raged on, I went on more than 400 combat patrols and survived 11 IED strikes, a couple of ambushes, and attacks on the FOB or PB we occupied. As often as I went outside the wire, the unit I attached myself to did it far more than I. Their strength of character, endurance, and professionalism was a testament to their mental and emotional readiness. As a result, they were able to seamlessly shift between two roles: on some streets I would see a Marine on one knee handing little children a stuffed animal or a piece of fruit, and then later, on another street, see the same Marine conducting sensitive site exploitation, gathering critical forensics following a shooting, or identifying unexploded ordnance meant to be used in an IED tomorrow against him.

I credit Joe L'Etoile with teaching me how focus, effort, and teamwork are essential to readiness. In the third year of the war in Afghanistan and Iraq, Lieutenant Colonel L'Etoile became battalion commander and I became the battalion sergeant major. We were sitting together one night during the workup, getting ready to deploy. In 60 days we'd be going back to Iraq, and we were rehearsing for every kind of event that could go wrong.

"Help me finish this list," he said, handing me a piece of paper. He had written down more than half a dozen potential dangers. "What's the probability that these things are going to occur when we walk outside the wire? What else could happen?"

We came up with a list of things that could happen every time our Marines walked outside the wire, and grouped them into 11 categories: weapons, vehicles, first aid, SSE,[10] communications, language skills, cultural knowledge, tactical SOPs, hasty/fixed positions, ROE, and CIED.[11] We needed to accomplish more than a thousand infantry training standards, missions, and mission-essential tasks and exercises during our six-month work-up, but these were the 11 things we needed to focus on to bring everyone safely home. We developed a master's degree program and focused on those skill sets.

During our deployment to Iraq, our battalion had more success in finding the enemy and killing them than any other battalion. Success and leadership did not motivate us; we were motivated because we knew this master's program would save lives. Good habits and disciplined warfighters—along with aggressive and highly trained tactics, techniques, procedures, and methods—drove positive strategic actions. Joe L'Etoile taught me that readiness requires all your focus and effort. Keep your head on a swivel, take care of that

---

9 High-value target
10 Sensitive site exploitation
11 Counter-IED

Marine on your left and right, and when it's time, shoot straight. Prepare as a team.

Lieutenant Colonel L'Etoile and I did two back-to-back deployments to Iraq as a team, and we went on more than 300 combat patrols together. Today, he serves the Secretary of Defense in training and readiness across the DOD.

Personal and Family Fitness

Are your personal affairs in order? Are your bills paid, living will current, shots up to date, direct deposit squared away, and your car running the way it's supposed to? Are your annual requirements current? Are you a rifle and pistol expert? Are you exceeding your professional education requirements? Are you a black belt in martial arts? You cannot be expected or entrusted to take care of others if you're not taking care of yourself.

Are your direct-deposit allotments sufficient to sustain the kids and the bills? Does your family know your duty, field, and deployment schedules, when and where you're going, how long you'll be gone, when you're leaving, and when you're expected back? If you have little children, have you made books on tapes or videos, enough to get through the deployment? It's important for them to hear you and see your face. Is your family linked in to all the family support groups and programs to assist while you're away?

Is your family socially fit? Do they have friends or relatives close by, prepared to jump into action if the need arises? Is your family spiritually fit? Are their hearts happy? Are their minds in a good place? Are they resilient? Are they engaged in activities that allow them to grow and develop?

Personal and family fitness is an absolute necessity. A Marine whose head is not 100 percent in the moment, on the task or mission at hand, can be a liability. A Marine on patrol in the most dangerous place on the planet, or a mechanic turning a wrench on an aircraft or MRAP, cannot perform as needed if things back home are not right. If their thoughts and concerns are on the broken car at home or the wife who isn't getting the money she needs, they won't be looking for the IED two steps away.

Moral Fitness

The Rifleman's Creed that we recited daily in boot camp included this principle: "I must master [my rifle] as I must master my life . . . I will keep my rifle clean and ready, even as I am clean and ready."

Moral fitness is an imperative for an effective leader.

"Treat your body like a shrine, not an amusement park," I would counsel my Marines. "I want you to live your life aggressively, not recklessly. A night of passion could result in a lifetime of pain."

Spiritual well-being is imperative for moral fitness. When I say "spiritual," I'm referring to what's in your heart, not necessarily religion. "Spiritual wholeness" means all parts of you are well and working together. Mind, body, and spirit make up the whole Marine. ■

Lesson Five: Give Your All

I am well aware of the toil and blood and treasure it will cost to support and defend these States. Yet through all the gloom, I can see the rays of light and glory.
~ John Quincy Adams

I have sat and stood with presidents, members of Congress, ambassadors, Cabinet secretaries and other world leaders, all of them impressive in their accomplishments. But none has impressed me more than a 19-year-old lance corporal putting it all out there, leaving it all on the battlefield.

In June 2010, Cpl. Clifford Wooldridge was assigned to 3rd Battalion, 7th Marines when his company came under intense machine gun fire in Helmand Province, Afghanistan. After the Marine vehicles returned fire, dismounts were pushed out to close with the insurgents. Once the dismounts moved away from the vehicles, the enemy prepared for an ambush.

Corporal Wooldridge took his fire team and pushed around the suspected enemy flank, receiving fire from a tree line 100 meters away. He directed his Marines to return fire and close on the tree line, and once they arrived at the tree line they saw approximately 15 enemy fighters carrying machine guns, rocket-propelled grenades, and rifles, preparing to ambush the other portion of Wooldridge's dismounted squad. The fire team killed five enemy fighters and wounded three; the remaining fighters ran away behind the compound.

While his Marines held observation on the downed enemy fighters and waited to link up with the rest of the squad, Corporal Wooldridge picked up flank security. He heard voices around the corner of an adjacent wall and saw two enemy fighters moving into an ambush position less than 25 meters away. He immediately engaged with two bursts from his M249 Squad Automatic Weapon, killing both fighters. As he began to re-load his weapon, he noticed the barrel of a machine gun appear around the corner of a wall less than five feet away from him.

Out of ammunition, Corporal Wooldridge grabbed the barrel of the enemy's weapon, threw the fighter to the ground, and engaged in hand-to-hand combat. When the fighter tried to pull the pin on one of the grenades attached to Wooldridge's vest, he grabbed the man's machine gun and butt-stroked him to death. Corporal Wooldridge and his team saved the lives of their fellow Marines by dealing the enemy a tremendous defeat and instilling fear in the remaining fighters.

Corporal Wooldridge had the ability to remain calm and unruffled in this intense life-or-death situation because he was prepared to give his all. Holding back was never an option; he had already made the decision when he became a Marine to give 100 percent effort and focus to the fight. Nothing but death would have stopped him from defending his men.

While I was serving as Sergeant Major of the Marine Corps, two requests to redeploy came to my desk. Capt. Matt Lampert, MARSOC,[12] a double-leg amputee, and Sgt. Jason

---

[12] Marine Corps Forces Special Operations Command

Pacheco, sniper, a right-leg amputee, had directly contacted General Amos and asked to redeploy to combat in Afghanistan. There's no way I could deny the request of Marines who wanted to sacrifice a safe and comfortable duty for the real warfighting mission of a Marine. It was easy to say yes without hesitation. You don't become a Marine and hope to sit on the bench. Captain Lampert and Sergeant Pacheco were unreservedly committed to giving everything they had to fight alongside their brothers.

As a Marine, I trained hard at every rank, every day. My mind-set was "the day is over when the job is done," and I never settled for just "OK." I learned that the more I gave, the more my leaders gave back. I would volunteer for everything because I was afraid I would miss something important if I didn't. Just showing up on time in a squared-away uniform and doing what you're directed to do is never enough. *Every* day is qualification day!

You don't become a leader by doing just the bare minimum. Completing just five MCI[13] courses so you can have additional points for a cutting score is about five short of what you should be doing. Running a first-class PFT[14] is not the same thing as running a perfect 300. You want to earn an expert marksmanship badge, but if you didn't spend time every day after the range with that Marine on your right or left who was having trouble, then you missed the mark. Reading three books a year from the CMC's reading list is good, but it's the low end of the mark. That's just the suggested number. As soon as you finish a book, pick up a new one.

I not only use all the brains that I have, but all that I can borrow.
~ Woodrow Wilson

Whatever your specialty, whatever the task you've been assigned, bloom where you're planted. Do more than is required; go above and beyond. Make sure you have the knowledge and skills required not just to perform your current job, but also to assume the duties and responsibilities of the next grade or next-higher billet. If you are the sergeant major of a battalion, can you jump on the .50 cal and get it into action or clear a misfire?

When I was Sergeant Major of the Marine Corps, we had a saying: "That's why God gave us 29 hours a day, and 9 days in every week." A day never went by where we didn't learn something new from each other. There was always time every day to share our strengths with each other and put all our effort and focus into team building. Doing anything less was not an option.

Go the extra mile. Open your eyeballs and turn up your hearing aids, try everything, do great deeds and endure. Sometimes you get the wind knocked out of you, but never quit.

I'll take attitude over aptitude any day of the week. (Of course, I prefer both.) Be positive and optimistic in all endeavors—it's contagious. Some fights you win, some you don't. Don't get discouraged. People succeed because they are resolute in endurance and attitude.

I don't lose any sleep at night over the potential for failure.
I cannot even spell the word.
~ Gen. James Mattis

---

[13] Marine Corps Institute
[14] Physical fitness test

Leaders don't complain. Live hard, endure pain, and when bad things happen, take them on! Ensure corrective actions are being taken. Educate. Establish and implement goals and policies. Determine if it was poor leadership, poor planning, poor organization, or inability to anticipate potential problems. Follow up on the corrective actions you've implemented.

Effective leaders accept responsibility for failure. Don't pass blame. Ask for forgiveness; explain what you were trying to accomplish. Demonstrate you were trying a new approach: you set an ambitious goal, and you were being innovative and creative. Talk about what, how, and where you failed, and what it taught you. Support the team to keep on reaching further than their grasp in order to accomplish great things. A positive mind-set doesn't ensure success, but a bad one makes certain a disappointing ending.

When crises occur, maintain your equanimity. You're not alone. Who are your talented fixers? They have already demonstrated multiple times at many levels their abilities to get things back on track. They are not unfamiliar with chaos, friction, or loss.

Life as a Marine is serious business when you're on duty or outside the wire, but don't forget to enjoy every other free second of the day. Live it aggressively with passion and with people you love to be around. Whatever you do, do it with vigor, enthusiasm, passion, and grit. Keep your wit and sense of humor.

I am afraid of only one thing: not trying. That is simply not an option. We will never get to where we want to be by remaining where we are. Take risks and take a stand. Lead the charge!

*   *   *   *   *

In my last week as Sergeant Major of the Marine Corps, I was returning from a meeting, and as I negotiated the 22 miles of Pentagon passageways, I came upon a guided tour. The docent was a young Soldier in his blues.

He introduced me, gave a brief description of my position, and asked if I had a minute to answer questions. I was happy to.

A young lady in her mid-20s asked, "How long have you been a Marine?" I told her I had served for 34 years. The young man next to her followed up immediately.

"I thought you could retire at 20 years," he said.

"You can," I responded.

"So why do you still serve?" he asked.

It was the thousandth time I had been asked that question, and a Sherlock Holmes story explains how I felt about it. Holmes and Dr. Watson went on a camping trip, set up their tent, and fell asleep. Some hours later, Holmes woke up his faithful friend.

"Watson, look up at the sky and tell me what you see," he asked.

"I see millions of stars," Watson replied.

"Well, what does that tell you?" Holmes persisted.

Watson pondered for a moment, searching for a profound explanation.

"Astronomically speaking, it tells me that there are millions of galaxies and potentially billions of planets," he said. "Astrologically, it tells me that Saturn is in Leo. Timewise, it appears to be approximately a quarter past three. Theologically, it's evident the Lord is all-powerful and we are small and insignificant, and it appears we will have a beautiful day tomorrow. And what does it tell you?"

"Watson, you idiot, someone has stolen our tent," Sherlock Holmes said.

I didn't respond to our visitor's question with the exasperation Holmes showed Watson, but I made it clear that the answer should be obvious to any patriotic American.

I described how I felt on September 11, 2001, when the most hateful men on the planet took the lives of 2,977 innocent human beings—mostly Americans, but also men and women from 90 other countries. I told them that only .4 percent of 320 million Americans serve in the U.S. Armed Forces, and that at no other time in our nation's young history have we been engaged in direct actions with our enemies for this long, with an all-volunteer force taking up the call.

"Most of our men and women aren't old enough to buy beer," I said. "They could all be doing something else with their lives, but instead they chose to wear the uniform to serve free people who live in a free country. Those of us who serve this country feel every note of the national anthem while we stand at attention with our hands over our hearts or rendering a salute. We don't bend to outside distractions—we even temper the insane desire to smack any person who fails to show respect to the flag."

The visitors laughed, but I was serious.

"We're an elite class of Americans that has kept this country free for over two centuries. We live hard, train hard, and fight hard, and we provide our nation with the capability to contain any crisis. We know what it means to keep company with the finest men and women in the world, under the toughest conditions. We live our lives right and to the fullest. We live by a set of values: honor, courage, commitment."

Based on the young man's enthusiastic nodding and wide-open eyes, I believe he now understood why I chose to keep serving my country, long after the obligatory 20 years were up.

Every day I was in uniform, I carried a pocket-size Declaration of Independence and U.S. Constitution in my left breast pocket to remind me why I was doing what I was doing. The profound words in those documents laid the foundation of what made me a better and stronger servant leader.

One of the reasons I stayed in the Marines as long as I did was because I know what people are truly willing to sacrifice for an idea: free people living in a free society, coming and going as they please. Nothing is stronger than the heart of a volunteer! ■

Lesson Six: Transitioning

March on. Do not tarry. To go forward is to move toward perfection.
March on, and fear not the thorns, or the sharp stones on life's path.
~ Khalil Gibran

I have nothing earth-shattering to pass on regarding transitioning from military life
to the private sector. As I stated in my introduction, "There's no magic in what I
did." I merely followed instructions, and my family and I moved seamlessly to a great
neighborhood where kind people take the time to wave and ask if you need anything.

I can't recall any challenges in my smooth transition from 34 years of military life to
that of a retired Marine veteran and civilian, excited to serve my new community. (There
was the eerie quiet of all my electronic devices falling silent, but that was a good thing.)
Why was the shift so smooth for Susan and me? The simple answer is that I took the
Executive Transition Readiness seminar at Camp Lejeune, and I followed the instructions
and recommendations of the incredibly caring and committed staff of our Transition
Readiness team. I recommend going twice—*and take your spouse.* The Marine Corps
Community Service team bends over backwards to ensure that all our outgoing service
members are best prepared for this life-changing milestone.

A few factors in particular made my transition almost effortless:

1. Get a final physical
My transition to VA care was excellent because I started my final physical with sufficient
time to have all my aches and issues covered prior to retiring.

2. Meet with your CVSO
Meet with your County Veteran Service Officer (CVSO) within the first month of taking
off the cloth. You're entering a new and foreign world. For me, my entire adult life was in
uniform as an active-duty U.S. Marine. I thought I knew everything I needed to—but I had
never retired before. I learned more about veterans' benefits in one visit to my CVSO than
I had learned in 34 years: healthcare, service-connected disabilities, home loans, burial and
memorials, property tax deductions, individual state veterans' benefits, etc. It will be an
afternoon well spent.

3. Stay connected
Stay engaged and connected—join the American Legion, VFW, or any of our wonderful
Veteran service organizations. You all have something in common: an amazing support
network. Whether you need help cutting down a tree, your car needs fixing, you want a
place to go shooting, you're negotiating red tape associated with the VA, or you're just
looking for an inexpensive beer, volunteers are there for everything and anything. Scores of
friends in these organizations are ready to leap into action if you ever find yourself running
low on motivation, or just need assistance.

Once you've been retired for a while, you can volunteer to be the one helping out.

4. Be involved in your community
Attend and celebrate every holiday and special occasion in your new community with family and friends. Thomas Robbins, a chaplain, wrote in 1822 about the reason societies celebrate certain occasions of importance:

> The manners and character of a community are essential to national prosperity. [...] He who loves his family and relatives, who possesses a disposition of charity and benevolence towards his neighbors and acquaintances, who conducts with integrity and kindness in his ordinary intercourse with his fellow-men, [...] in short, who seeks to "do good to all as he has opportunity," is a patriot, is a friend to his country.[15]

5. Share your story
Say yes when people ask you to speak. You have a story to tell and lessons to teach, and people want and need to hear about what we do and what we've done, the sacrifices and the victories.

6. Continue to serve
I still serve! I just wear a different uniform. The Corps has provided me with a remarkable skill set, and I will continue to give as long as I am physically and cognitively able. It's a privilege to have the opportunity to still provide a service. Stay fit, stay in motion, stay relevant, and keep serving.

*　*　*　*　*

Today I serve as a military advisor to Veterans United Home Loans (VUHL), an organization that has a positive impact on the lives of those who have served—or are still serving—their country. The company's ethos is engrained in all who work there; they're passionate about what they do, deliver results with integrity, and enhance lives every day.

I volunteer my time on two boards of directors. One is the Marine Corps Scholarship Foundation, a nonprofit organization that honors our Marines by educating their children. Since 1962, the MCSF has provided more than 40,000 scholarship awards worth nearly $120 million to children of Marines and Navy Corpsmen attending accredited colleges, universities, and career/technical institutions, with particular attention given to those whose parent has been killed or wounded in combat, or who have demonstrated financial need.

The second board is for the Young Marine National Foundation, a nonprofit that strengthens the lives of young people. They are the example of what "right" looks like: they are self-confident, work hard in school and achieve academically, serve their community as good citizens, live a healthy, drug-free lifestyle, and respect and honor authority, their

---

[15] Thomas Robbins, *An Address Delivered Before a Number of Military Companies: Assembled at Hartford, to Celebrate Our National Independence, July 4, 1822* (Hartford: Goodsell & Wells, 1822), 8.

family, veterans, and peers. The Young Marines program focuses on character building and leadership, and promotes a lifestyle that contributes to being a positive, constructive member of society.

I also volunteer my time as an ambassador-at-large for the Marine Corps League, a Veterans Service Organization, and support the Armed Services YMCA. Members of the Marine Corps League join together in camaraderie and fellowship for the purpose of preserving the traditions and promoting the interests of the U.S. Marine Corps. The ASYMCA is a nonprofit that serves active-duty military members and their families. Whether providing respite child care for parents in need, summer camps for kids, or assisting with emergency needs, the ASYMCA mission is to focus on the resiliency of military families.

I am extremely privileged to stand shoulder to shoulder with Ken, Rick, Jim, Denise, and Skip, my partners in Summit Six LLC. They are my comrades-in-arms, but more importantly, they are my friends. ■

TOP LEFT: Delivering the opening remarks, U.S. Chamber of Commerce, Washington, D.C., Hiring our Heroes Expo, 2014.
TOP RIGHT: Reading scripture at the Washington National Cathedral, U.S. Marine Corps birthday, 2011.
RIGHT: Susan and I enjoying friends at Sgt. Dakota Meyer's Medal of Honor ceremony, Marine Barracks, Washington, D.C., 2012.

TOP: Partnered patrol, Nahri Sarraj, Helmand Province, Afghanistan, 2010.
ABOVE: Atmospherics patrol, Marjah market, Helmand Province, Afghanistan, 2010.
LEFT: Talking with Marines, 7th Marine Regiment, Motor Transport Pool, Marine Corps Air Ground Combat Center, Twentynine Palms, California, 2014.

TOP: Sharing a meal with Marines at Camp Pendleton, California, 2013.
ABOVE: Susan and I hosting Lance Cpl. Justin Rokohl, Cpl. Anthony Villareal, and Lance Cpl. Ray Hennigar at a reception in the commandant's garden before the Friday Evening Parade, Marine Barracks, Washington, D.C., 2011.
RIGHT: Running a shoot (multiple gun) course at MCCDC, Quantico, Virginia, 2011.

TOP: Susan and I greeting guests at a reception, Truman Crawford Hall, Marine Barracks, Washington, D.C., 2011.
ABOVE: Susan visiting Marines at Marine Corps Air Station Iwakuni, Japan, 2011.
LEFT: Outside unnamed temporary patrol base, Musa Qaleh, Afghanistan, 2010. (In small print: *Violators will be shot! In accordance with ROEs, of course!*)

Rick D. West

Rick D. West became the 12th Master Chief Petty Officer of the Navy (MCPON) on December 12, 2008, and served until September 28, 2012.

He was born in Rising Fawn, Georgia, and entered the U.S. Navy immediately upon graduating from high school in 1981. After recruit training and quartermaster training in Orlando, Florida, he attended Basic Enlisted Submarine School (Naval Submarine Base New London) in Groton, Connecticut.

His first duty assignment was aboard USS *Ethan Allen* (SSN 608), where he completed submarine qualifications. Other assignments included USS *Thomas Edison* (SSN 610), USS *Sea Devil* (SSN 664), Commander Naval Activities United Kingdom (COMNAVACTUK), USS *Tecumseh* (SSBN 628)(Blue), and Commander, Submarine Force, U.S. Pacific Fleet (COMSUBPAC) Staff (TRE Team).

West was assigned as chief of the boat (COB) aboard the San Diego-based fast-attack submarine USS *Portsmouth* (SSN 707), where he completed two Western Pacific deployments and was selected to the Command Master Chief (CMC) program. The *Portsmouth* crew twice earned the Battle Efficiency ("E") Award.

Upon completion of a CMC tour at Submarine Squadron (COMSUBRON) ELEVEN, West was selected as COMSUBPAC force master chief (FORCM) from 2001 to 2004. During this time, he also attended the Senior Enlisted Academy (SEA) in Newport, Rhode Island. After completing the force master chief tour, FORCM West then reverted back to command master chief and reported as the CMC to USS *Preble* (DDG 88) in San Diego. He deployed on the ship's maiden voyage to the Arabian Gulf and qualified as an enlisted surface warfare specialist. The crew earned the *Preble*'s first-ever Battle Efficiency ("E") Award.

West was selected during his tour on *Preble* to serve as Pacific Fleet master chief (FLTCM) from February 2005 to June 2007. Following PACFLT, he served as fleet master chief for Commander, U.S. Fleet Forces Command from June 2007 to December 2008. West was the first senior enlisted leader to serve as fleet master chief for both operational fleets.

West's personal awards include the Navy Distinguished Service Medal, Legion of Merit (two awards), Meritorious Service Medal (three awards), Navy Commendation Medal (four awards), Navy Achievement Medal (two awards), Enlisted Submarine Insignia, Enlisted Surface Warfare Insignia, and SSBN Deterrent Patrol Pin.

He retired from active duty in January 2013 and currently works for Progeny Systems. He also serves as a military advisor for Veterans United, is active within the Navy supporting boards and Navy chief petty officers community, regularly speaks at Navy events, and conducts fleet visits.

West and his wife, Bobbi, have two sons and live in Poulsbo, Washington. ∎

Introduction: A Strong Work Ethic

Without hard work, nothing grows but weeds.
~ Gordon B. Hinckley

I love being a Navy Sailor—I wasn't born one, but I sure as hell will die one. My life began about 500 miles from the ocean, in the rolling hills of northwest Georgia. Nestled between Fox, Lookout, and Sand Mountains is a podunk town of around 500 people called Rising Fawn. I love the people, love the town, love the name. It was a great honor to rise through the ranks and become Master Chief Petty Officer of the Navy (MCPON), but my journey to get there started long before, as a kid growing up in Rising Fawn.

To understand my style of leadership, you must first know where I came from and what shaped me into the man I am today.

When I was a kid, my parents' service station and dairy bar just outside town was a gathering spot and local hangout—some called it "the BS place," since you could always count on a tall tale or two as people clustered around the Coke machine. Let's just say the banter I heard helped me grow up. The place initially did good business in the 1960s while Interstate 59 was under construction, and it was continually bustling with activity.

It was a busy time for all of us—my parents certainly couldn't do it alone and we couldn't hire much help, so the three young West kids picked up the slack. Our parents taught us how to work hard and gave us big responsibilities to support the family business. (In retrospect, I would say we worked like dogs.) I learned early responsibility by observing my parents' work ethic, and they demanded the same efforts of Johnny and Brenda and me.

By the time I was six, I had a grease rag in my pocket and was responsible for light vehicle maintenance, such as checking tire pressure and oil levels. On occasion, I was even allowed to make the swirly ice cream cones at the dairy bar, flip a burger, or pump gas when we had a surge of customers coming in at one time.

This was a full-service station—not something you see much anymore. The routine looked something like this: several cars would pull in for fueling, and my brother and I would immediately stop playing sports and come running across the lot to help the customers.

This is where my "customer service" attitude began. My parents demanded high standards in whatever we did, and we always had to treat customers with the utmost respect. "Yes, Sir," "No, Sir," "Yes, Ma'am," and "No, Ma'am" were a must, and we would always get "the look" if we even *considered* interrupting a conversation between adults.

One of the highlights of my day was when the work crews on the interstate would stop by to order lunch from the dairy bar, then sprawl out under the pine trees and share their thoughts on the world. Thanks to my mom and sister's great cooking, people would drive for miles to eat their cheeseburgers, fries, and milkshakes. The road crews were groups of men and women from all walks of life, and I quickly made friends with them. They would pick me up when I was down, and they provided a running and colorful commentary on life that my mom didn't always approve of. That didn't stop me from hanging out or throwing a baseball around with them, but I didn't take part much in the conversations. As a kid, my job was just to listen. We didn't interrupt adults, and we only spoke when spoken to. My dad always said that a wise man should listen and learn.

Dad was tough and a man of few words, so when he spoke, we listened. During my childhood, he drilled certain principles into my head, and I applied them to be successful in whatever I did. Whether it was on a ball field, wrestling mat, submarine, or ship, I followed Dad's example of leadership, values, and attitude. These principles formed the foundation of my life.

Dad's principles
1. Never be late.
2. Work hard every day. Nothing in life is free; earn your keep.
3. Stay out of trouble.
4. Respect everyone, especially your elders.
5. Don't make fun of anyone, especially the less fortunate.
6. Finish what you start.

Dad was a man of ideas—some good, some bad—and both he and Mom worked their butts off to make ends meet. We lived in a small house of less than 1,000 square feet, and the usual dinner consisted of beans, taters, and corn bread. Although we owned our home, Dad rented acreage, and as I grew up we acquired a motley assortment of cows, horses, chickens, and pigs, among many other darn things. It was not uncommon back then to have people pull up to get gas and not have the money to fill up their tank, so Dad would offer them credit. He had a stack of 3x5 "IOU" cards about as tall as the cash register. Sometimes you might see a customer unloading a goat, guns, or vegetables to finish the payment, but many folks didn't repay their bills, or they just skipped town without a word. Dad's well-intended generosity likely led to the demise of our business.

We grew much of the food we consumed. In the summer, we ate fruit and vegetables fresh from our garden, and we canned plenty of it for the winter. Because we kept livestock, we always had beef and pork in the freezer. When I was about 11 or 12, I would occasionally pack a backpack and my shotgun or fishing rod and head off into the woods for a day of hunting, then come back with squirrel, deer, rabbit, or fish to add to our stores. Dad would look for slow times at the station during the day, then quickly load me up in our 1966 VW Bug and take me to help him feed the animals and work the garden, just a five-minute drive away.

Dad believed our garden should feed the town and any others who came by the station. Today the size may not seem that daunting—although it was large enough to require a tractor—but to a young kid, it was the biggest damn garden in the world. That work never ended. I still break out in a cold sweat when I remember having to hoe what seemed like endless rows of corn in the steamy Georgia heat, all the while keeping an eye out for the occasional rattlesnake or copperhead lurking around.

I could go on for hours about the hard work my parents expected of us, and the resulting work ethic deeply ingrained in me. To this day, I take great satisfaction from putting in a long, hard day of work. The flip side to that, according to my wife, Bobbi, is that I don't have an "off" switch; I find it hard to just relax and spend a day doing nothing. You do need balance in life, and I'm working on it!

But growing up in Rising Fawn wasn't all work, even though that came first. Mom and Dad bought me a go-cart when I was about eight, and in my eyes, I could have rivaled Richard Petty and Cale Yarborough on the racetrack as I zoomed helmetless around the neighborhood and gas station, often swerving to avoid getting hit by customers and then avoiding my mom's wrath after witnessing the near miss. It wasn't just for fun, though. Driving that go-cart taught me responsibility and gave me confidence—possibly too much.

When I was about eight, I learned to drive my Uncle Bill's truck. I was the driver when we worked in the hay fields, since I was too little to be much help throwing hay. Driving a clutch in a hay field is my first memory of being truly humbled—after throwing a few of the guys off the back of the truck and dumping the hay, I finally got the hang of it. By the age of nine, I was lifting the hay bales. Just the other day, I saw a load of hay on a flatbed truck, and the sight put a huge smile on my face. I believe every youngster should have to work in a hay field at least once in Georgia heat—everything they deem "hard" will pale in comparison.

On Sundays, we often took a family drive on one of the dirt roads. One particular Sunday when I was about nine or ten, my dad and I were out on a leisurely drive when he decided I needed another driving lesson.

"Son," he said, pulling the car over, "you have to learn how to drive a car on the highway at some point, and now is as good a time as any."

Delighted, I got in the driver's seat. We were about 45 minutes into the drive on the back roads, and Dad was two beers down in a six-pack of Schlitz, when a sheriff's deputy whizzed past me around a blind corner. In my opinion, he was going a bit too fast. From the deputy's perspective, he probably thought he saw the very first self-driving car, because I could barely reach the pedals or see over the dash.

I remember looking in the mirror and seeing brake lights.

Uh oh, I thought.

My dad, thinking quickly, tossed the four full beers out the window before the deputy turned around and headed back to us in hot pursuit. Dad's biggest job at that point was to settle me down as I burst into tears, thinking I was penitentiary bound.

I kept begging Dad to switch seats with me, but he said, "Son, what's done is done, and I'll be the one in trouble—not you." The deputy caught up with us, lights flashing, and I was about to have my first real brush with the law.

I pulled over, and Dad told me to stay in the car while he talked to the deputy. I was sure I was headed to San Quentin Prison, which I had heard about from listening to Johnny Cash's protest song, "Man in Black."

"Mr. West, I know that boy is not 15," I heard the deputy tell my dad. That was the age for a learner's permit.

"No, Sir," my dad replied. "He's not, but he's got to learn responsibility and how to drive at some point."

"You're the West who runs the gas station on the hill?" the deputy asked, looking at my dad's license. Dad nodded, and the deputy handed him back his license.

"Look," he said, "be careful. Don't let him drive on the main roads where the troopers are."

He let us off with a warning, and we proceeded with the rest of our Sunday drive—with me in the passenger seat, humbled but not broken. I learned a huge lesson that day. I didn't think I was wrong in driving, but I wanted to change seats before we were busted. Dad told me that switching seats wouldn't have been honest; we needed to accept responsibility for our actions. To my dad, there was a clear line between right and wrong.

Other folks helped me along the way, too, and gave me tools for the success I've had. Though I graduated high school early and left Rising Fawn at 17 to join the Navy, many teachers were my mentors, and I honestly could name almost all of them today.

My elementary school teachers Mrs. Street and Mrs. Hawkins tolerated and challenged me. Mrs. Rose Moore, the librarian throughout middle school, was always there to provide a firm but guiding hand. So many others shaped me as I was growing up. These men and women don't get paid enough, yet they take the burden of influencing and educating our youth. I'll always remember them and wish I could thank them for the journey. I certainly wasn't the best student, but I held my own in the classroom. On the other hand, I don't think I deserved the paddlings I took for being a knucklehead or popping off at the mouth! But like it or not, that was part of my journey, and I'll always look back fondly on the times I got in trouble. I learned something each time.

*   *   *   *   *

When I was 13, I wrote the Army recruiter in nearby Chattanooga, Tennessee, and told him I wanted to be the best darn Soldier they'd ever seen. I think it was all the John Wayne movies I'd watched. To my surprise, the lead recruiter wrote me a nice letter back, telling me I had to be older and they would be there for me when I was ready. (*Didn't they know I could drive already?* I remember thinking.)

During the next few years, while I waited to turn 17, I decided the Navy was the best place for me. As a young boy, I had welcomed my cousin Morris home from the Vietnam War. He had deployed on the aircraft carrier USS *Kitty Hawk* (CV 63), and seeing him looking proud in his Navy crackerjacks made a positive impression on me. The stories he told and the places he visited made me realize a bigger world was out there, and I began to want and need to be part of that.

A few years later, I went to work at the local KOA campground when I was 16. Jim and Merlene Meyer ran the place. Talk about high standards and discipline! If you ever stayed at that little campground nestled in the Georgia foothills, you would know what I mean. The Meyers ran a tight show and put customer service and quality of life at the forefront. I'll never forget how humbled I was after my first assignment to clean the bathrooms. I broke out the gear and put several hours of effort into that task, and I was quite proud of myself and happy with the results. Both Mr. and Mrs. Meyer then inspected my work. I stood by with the toilet brush in my hand, expecting them to praise me. Instead, Mrs. Meyer began to point out deficiency after deficiency, taking me down a notch with each defect in my work. My standards didn't quite match theirs.

But then she and Mr. Meyer both grabbed a brush and took the time to show me exactly what their high standards were. Honestly, I went from being a little mad… to humbled… to being competitive enough to not let that happen ever again. I had never had a leader "inspect what they expect." For the managers of this campground to take the time to show me what they expected left a mark on me forever. As the Irish proverb goes, "An ounce of example is often worth a ton of advice." I often used this experience during my time in the Navy. "Lead by example" was a concept I learned early and used often.

You Sailors in the Navy who know my cleaning standards, you can thank Jim and Merlene Meyer for that. Mr. Meyer often talked about his experience as a commissary steward in the Navy, and his positive experience, high standards, and sea stories significantly influenced and sealed my decision to join the Navy. Sir, I thank you for taking the time to motivate, challenge, and push me in that direction.

Something else the Meyers taught me was the benefit of getting out of my comfort zone. Knowing I could use the money, Mr. Meyers asked if I would assist in giving horseback rides after my normal hours of work. Although I had grown up around horses, I had little riding experience—I might have exaggerated my horse-riding skills a bit. I was assigned as the lead guide, but after a few rides I quickly realized I was way out of my league. The horses had a mind of their own, and one day the group I was guiding scattered all up and down the trail. Luckily, no one was hurt. After that near miss, I determined to improve my skill and knowledge level. Let's just say the rides were all successful from that point on!

About the same time I was working at the KOA, the Navy launched some cool TV commercials with the tagline "It's not just a job, it's an adventure." They really caught my attention and made me believe that even a small-town Georgia boy who had never seen the ocean could sail the seas and travel the world. All those influences led me to join the world's greatest Navy in 1980. Since I was only 17, my parents had to sign their consent. After much discussion with Dad, Mom finally relented, and off I went to the recruiting office in Chattanooga. Just like today, all the services are typically represented at the same place; you walk down the hall and have your choice of which one to join. By this time, the commercials and the influencers had me 100 percent Navy. Still, I remember walking past the Marine office, and one of the Marine recruiters dropped his pencil on the floor. I stopped to pick it up.

"Son, see that pencil?" he asked.

"Yes, Sir," I responded, handing it to him.

"That pencil is small on this big floor," he told me, "just like a ship on the ocean."

Although I applauded his tactics, I handed him the pencil and continued my journey down the hallway to the Navy office. Sailing the world's oceans and seeing sunrises, sunsets, and foreign ports really appealed to me.

After a series of tests to determine what I would be qualified for, I sat down with the recruiter to determine my Navy path. Like many young men and women today who join the Navy, I had no idea what I wanted to do or what was even available. I asked him what he did. He was a senior chief aviation electrician's mate (AECS) and he loved it, so I signed the dotted line and was admitted to the Navy's Delayed Entry Program as an AE. Yes, Rick West signed up to be an aviator.

The rest of the summer was filled with playing baseball, having fun, and working at the KOA. I had just finished 11th grade, but had enough credits to graduate early. A high school buddy had also joined the Navy, and pretty soon I got letters telling of his exploits and how he loved being a radioman on a submarine. Between Tony Smith's letters, Mr. Meyer's guidance and boot in the stern, cousin Morris's stories, and Dad telling me I should see the world and do something bigger than myself, I was itching to go. I called the recruiter and asked if I could leave earlier and change to submarines as an undesignated fireman. A few hours later, I was sitting with my mom when he called back with an answer.

"You're shipping out tomorrow," he said briskly. "You've got about two hours to get your stuff ready for boot camp."

I'll never forget driving down the hill from our house in my VW Bug, looking back in the rearview mirror and seeing my mom crying on our front porch. I never knew how she felt until I experienced it myself when I was a dad.

*　*　*　*　*

I learned a lot of lessons, both good and bad, listening to the adults talk at the gas station about "how it used to be" and how my generation would mess the world up.

"Kids today just don't get it," they would say, shaking their heads. "They'll never understand what we had to go through and the struggles we had to endure. We had it a lot rougher than they will. Kids these days are just spoon-fed and soft!"

Any discussion like that always had the same ending: "I'm worried about what the world will look like in 20 or 30 years."

I always listened intently, wondering who and what they were talking about. I guess with all their concern about what the world would look like 20 years down the road, that's about when I developed a sense of adventure—I wanted to get out and see the world and be part of the generation that would mess it up.

Fast-forward almost 50 years later, and the same conversations are being held in small towns all across America. In fact, the #1 question I received as MCPON was, "Are you worried about the young Sailors today?"

My answer was and is, "Absolutely NOT."

These conversations will still be around long after I'm gone.

I'm not worried about the youth of today, although I may not agree with many of the things they do. I'm more worried about those like me, the leaders who haven't been exposed to America's best and may still think our future is somewhat grim. I, for one, think they're up for the challenge.

I do believe, however, that youth need to be guided in the right direction and given a sense of purpose. Bobbi and I did our best to teach our two sons what we were taught as youngsters: nothing is free, but if you work toward your goals and stay positive, good things will happen. ∎

Lesson One: The Navy Is a Team Sport

It follows then as certain as that night succeeds the day, that without a decisive naval force we can do nothing definitive, and with it, everything honorable and glorious.
~ George Washington

I believe that if you work hard, stay out of trouble, do the right thing, and finish strong, you'll succeed in life. What is success? It's defined by what you do for others, not for yourself. My best days were the days I did something that allowed even just one of my Sailors to succeed. Many times I looked at my Sailors as my kids, and there is nothing more special than seeing those you've raised or led be successful.

Life is good, people—live every day like it's your last! Smile often, be enthusiastic, find your HOOYAH, and give those in your life the chance and ability to do the same. Rarely will you find an opportunity, as military leadership offers, to shape hundreds and thousands of lives and push them towards success. I have a high tolerance for doers and those who have a positive spin on life, and a lower tolerance for the glass-half-empty crowd who looks for excuses and why they can't.

As far as failure goes—yes, it will happen, but learn from it, don't make the same mistake twice, and, most importantly, don't make a habit of it.

This book is not a "how to" manual for leadership, but more a view of some of the things that shaped us and led to our success as leaders. My hope is that you will pick at least a couple of principles from our experiences and use them in your day-to-day routine. Use them to first become a better follower, and then to become a better leader.

*   *   *   *   *

During my time on the Submarine Force, U.S. Pacific Fleet as a TRE[1] inspector and a Training Resource and Schools coordinator, I rode every submarine in the Pacific. My focus was to assess a sub's navigation department, but I also got to see how the team operated in a stressed environment. "TEAM" is sometimes used as an acronym for "together everyone achieves more," and that is certainly the case in a submarine. The Navy is not an individual sport—it's a team endeavor all the way.

One thing that makes the military in general (and the Navy specifically) such a great place is that young people from all walks of life learn how to work as teams to carry out our nation's bidding. Our Recruit Training Command in Great Lakes is an amazing place—the quarterdeck of our great Navy. Great Lakes receives a lot of feedback about how they're doing or what they're not doing, but they do a great job of getting Sailors ready for the fleet. Our responsibility as leaders is to ensure we continue challenging and helping them grow into great Sailors and citizens who represent not only our great Navy but also our nation. You never stop learning as a Sailor or a leader, whether you're the newest Sailor at

---

[1] Tactical readiness evaluation

the command, the CMC of a command, or the MCPON or CNO[2] of the Navy.

From the first day in boot camp, you learn the essentials: how to speak and act like a Sailor, and how to form a team. We have some of the best technology our taxpayers can buy, but it's the people who are the heart and soul of everything we do. Without our people and their ability to work as teams, all that technology, weapons systems, and platforms would just be hunks of metal. Our Sailors bring all that to life and are at the heart of mission accomplishment.

The moment I stepped into the Navy, I knew it was for me. Hell, blue-and-gold blood ran through my veins from day one, and it still does. I love the sounds, the smell, and everything about being on a ship or around a group of Sailors. In 1981, I reported to the newly converted slow-attack submarine USS *Ethan Allen* (SSN 608) in Bangor, Washington. There was no sponsor program,[3] so I had to find my way to the base. I was assigned a room and two roommates, but they weren't there and it was a weekend. I sat in my room that Saturday until I heard others, and finally ventured out to get something to eat. Someone was kind enough to point me in the right direction to the galley. On Monday, someone knocked on my door and took me to the boat for check-in. In 1981, of course, we didn't have cell phones or email; all information was sent via snail mail or naval message, or leadership told us what they wanted us to know at quarters. And there was no coddling in the Navy; married personnel were reminded that spouses didn't come in their sea bag and told to suck it up. Luckily, this mindset has changed.

Reporting aboard a naval vessel is an exciting but humbling experience. You're joining an established team—in my case, on a nuclear submarine—and you're judged the moment you walk on board. You have two ways to approach this: either put your head down and work hard to be part of the team, or not. You will be tested from day one, and the rest of the team will judge your attitude, work ethic, and desire to learn and support the command. They will observe something as simple as the way you qualify to stand a watch or the way you interact. The rest of the team is depending on you, so everything you do matters. This principle also applies to my time now in the civilian sector.

My first chief was QMCS[4] Gary Clinefelter—exactly the kind of guy you picture when you think of a Navy chief. He had the confidence, knowledge, and patience for everyone who approached him, and he was always there with a firm hand to guide us or a swift kick to our butts if we got out of line. He engaged early and often.

One of the first things Senior Chief did when I reported aboard was take the time to review my service record. His first advice and strong recommendations came on a Friday afternoon. I was in berthing cleaning up when I heard him call my name in his usual strained voice.

"West! West, where the hell are you?"

"Here I am, Senior," I replied, stepping into the passageway.

---

2 Chief of naval operations

3 Today, the Sponsorship and Indoctrination Program provides incoming Navy personnel with pre-arrival communication, adequate support upon arrival, and relocation assistance.

4 Senior chief quartermaster

"West, I reviewed your service record," he said. "I see you finished at the top of your class at Quartermaster "A" School and can be command-advanced."[5]

"I think so, Senior Chief," I replied hesitantly.

"Well don't just think so—you did!" he barked. "And the only thing you must do is extend for another year, do your PARs,[6] and then you'll be command-advanced to third class petty officer."

"Thanks, Senior, I'm not interested, but thank you," I told him, thinking it would be the end of the subject.

"Thanks for the input," he fired back. "Here's what's going to happen. Your liberty is secured for the weekend and you'll stay on board. We'll grade your PARs on Monday, and then you can sign the year extension and we'll look to advance you soon."

At that time in my life I was perfectly satisfied with staying where I was, but I did what any young Sailor is taught to do.

"Thanks, Senior Chief," I responded. "I look forward to the opportunity to excel and see the boat this weekend."

Although I didn't appreciate it at the time, I now know that action propelled my career in the Navy. I needed the push, and this senior chief gave me the rudder and direction. I stayed on board the submarine that weekend, completed my PARs, signed the year extension, and became QM3(SU)[7] West. Was it painful? You bet. But as I look back on it, it was one of the best things to happen to me as a young Sailor. I needed the guidance and got it; I needed the kick in the stern and got that, too.

A few weeks later as we were inbound from sea in the Puget Sound area, I was busting suds—doing my duty to the Navy by working in the galley. It's what Sailors call "mess cooking" or "cranking." I heard folks scrambling around the boat and felt a slight bump. A sailboat had suddenly come out of the fog, forcing the sub to make a quick maneuver in the restricted channel. Our stern hit the mud with a soft *thump*.

At the time, I thought my mission was washing dishes, but that was about to quickly change. The board of inquiry that investigated the accident determined that we had too few quartermasters (navigators), and they wanted to know why a QM was in the galley when the division needed me.

The next thing I knew, I was wiping off a table after serving the meal, and the chief of the boat[8] walked up to me.

"West, you're relieved from mess cooking duties," he said. "Go to Control."[9]

"Why?" I responded, curious. I quickly went from elation to getting my butt kicked for asking the COB "why."

After reporting back to Control, I was placed on a fast and furious pace to get my quartermaster qualifications completed. It was challenging, but having me fully qualified was

---

[5] COs can advance Sailors who are ready for the next level of responsibility, according to guidelines.
[6] Professional Advancement Requirements: correspondence courses that had to be completed to qualify for advancement within a rating.
[7] Quartermaster third class, in training for submarine qualifications·
[8] The chief of the boat, or COB, is the senior enlisted advisor to the commanding officer and executive officer on board a submarine.
[9] I completed my 124 days of mess cooking in port. The COB had the final say, and he got his pound of flesh.

necessary for our navigation team—they needed another QM to shorten the period of duty. Having me on the team was the key to getting them out of port and starboard watches[10]— six hours on, six hours off—sometimes back and forth for days, weeks, and even months.

The first time I stood watch, I'd never been more nervous in my life. It was now my job to ensure the submarine was properly navigated in a small operating area of the great Pacific Ocean. I don't even know how many pounds of sweat I lost that first six-hour period—I was constantly moving, checking, and rechecking our position. I couldn't have been happier when QM2(SS)[11] "Kermit" Crowley came around the corner to properly relieve me.

It was the most challenging six hours of my life, but in hindsight, one of the most rewarding. How the hell could the Navy give an 18-year-old young man from the hills of Georgia the responsibility to navigate and keep a submarine in a safe area? It's astounding, but this responsibility is given every day in our great Navy, and our Sailors deliver with outstanding results.

Thankfully, QMCS Clinefelter was the diving officer of the watch in the same section where I was assigned as quartermaster of the watch for a few weeks. He always had a watchful eye across the control room, should I need his assistance. He never got in my way, but I knew if I needed anything I could always ask and he would be there. He never needed to raise his voice; he had a chiefly presence about him that I respect to this day.

*　　*　　*　　*　　*

I made quartermaster second class on my first try, transferred to the converted slow-attack ballistic missile submarine USS *Thomas Edison* (SSBN 610), and immediately departed on a Western Pacific deployment in late 1982. Originally designed to carry Polaris missiles, the submarine was at that point in more of a "fast-attack" mode, doing conventional missions that set the stage for our SSGN[12] program today. I was quartermaster of the watch the day we reached our first port call. After the boat had surfaced and we rigged the bridge, we could see Yokosuka, Japan, in the distance. What a sight! I remember thinking how only 40 years prior, the United States and Japan had been at war. Our next few ports included Chinhae, South Korea; Phuket, Thailand; and the Philippines.

We took on some challenging operations. During one of them, an underway[13] out of the Philippines, we were the target boat operating against the nearly brand-new Spruance-class destroyer USS *Leftwich* (DD 984). These new destroyers were the pride of the surface fleet at the time and were supposed to be submarine hunters. We were giving them fits.

We had been causing havoc all night, launching flares and simulating attacks, when the captain gave the order to return to periscope depth to launch another simulated attack on the destroyer. When we started up, the pinging from the destroyer's active sonar,

---

10 "Port" and "starboard" refer to opposite sides of a ship. A "port and starboard watch" refers to a duty list that has two sailors alternating a watch station during a 24-hour period.
11 Quartermaster second class, submarine warfare specialist
12 The U.S. Navy's hull classification symbols for cruise missile submarines are SSG, SSBN, and SSGN. "SS" denotes submarine, "G" denotes guided missile, "B" denotes ballistic missile, and "N" denotes nuclear power.
13 "Underway" is a Navy term meaning a ship has pulled out of port and is out to sea.

looking for us, was so loud we had to cover our ears. It drummed against the hull, but we continued up, slow and steady. When the scope broke the water, the captain took about a half turn, then jerked away from the scope with eyes as big as saucers.

"Emergency deep!" he screamed.

In the next instant, we all heard the rush of water above and then a dull metal-on-metal sound as the *Leftwich* and the *Edison* collided. The impact jarred many of the small hull penetrations, which then began leaking. Add depth and pressure, and even a small leak looks major. Water started pouring into the submarine. Many Sailors that day thought it was their last ride, me included. I've told my family that I've feared for my life on a few occasions, and this was one.

We were driven under so hard that our fairwater planes, which protrude from the ship's sides and help control depth, were slammed hard; they flapped to the side, useless. Finally we arrested our descent and attempted to control the leaks.

This was my first real-life lesson on how crucial teamwork is on a submarine. The crew of the *Edison* could not have handled this any better. They executed as a team, assessed and repaired the damage as much as possible, and limped into Subic Bay. I had never seen so much khaki gathered in one place, waiting on the pier to receive us and go over the damage. The Cold War was at the forefront and the Soviets were aggressive, so leadership was alert.

We stayed in Subic for 35 days while they made the ship seaworthy for a 30-day surface transit back to the United States for repairs. A submarine is not built for a surface transit, so the suck meter was pegged high during this month-long journey.

The importance of having tough, high-standard warfare qualifications stuck with me from this experience. In the submarine force, it was mandatory for every crew member to qualify—it's an indispensable part of a Sailor's development and knowledge base. We saw, in a second, that knowledge can become the difference between life and death.

The significance of that idea stayed as I rose through the ranks, and when I became MCPON, I pushed the effort to make warfare qualifications mandatory across the Navy. Every Sailor needs to know how to "fight the ship," as it strengthens the team. Of course, qualifying is really just the beginning—learning is continual, and most of it comes after the qualification process. A couple of the force master chiefs did not want warfare qualifications to be mandatory; they worried that young Sailors in their force didn't have the time or the ability at that age to obtain them. Firmly believing we should never underestimate our young Sailors' brilliance and ability, I pushed to make the warfare qualifications mandatory. I'm confident that decision will strengthen our Navy and possibly save lives and ships. Our Sailors qualify for knowledge and train like they're in a fight.

*   *   *   *   *

As I moved up the ranks as a petty officer, I learned more about leadership and the ins and outs of not only managing Sailors, but also leading them. Managing and leading are two separate skills, and equally important. The ability to do both came as a result of increasingly challenging positions both ashore and at sea.

Although I enjoyed every submarine I was assigned to, the turning point for me came on board the fast-attack submarine USS *Sea Devil* (SSN 664), homeported in Charleston,

South Carolina. I reported on board as a somewhat senior second class petty officer early in 1984, having been fortunate enough to be command-advanced after Class "A" School. By then, I had served aboard two other submarines and had deployed on the aircraft carrier USS *Coral Sea* (CV 43) during her last deployment, which was my first experience with Big Navy and an aircraft carrier up close and personal. On that deployment, we were tasked to rendezvous with two other battle groups in the North Pacific Ocean and make a direct course to the Soviet Union. To our surprise, there was little to no reaction to our movement. Years later, I learned why: John Walker, the notorious sailor-turned-spy, had given the Soviets our plan.

The *Sea Devil* was a steamer, which meant we were gone a lot. During the Cold War, it wasn't uncommon to walk on board in the morning and have someone meet you topside with a change in plans.

"Go pack your bags," the COB would say. "We're going to sea, and I don't know when we'll be back."

Although the boat was frequently gone, the crew made the numerous underways pass quickly and set the tone for my naval career. Although I had served on some good submarines, the *Sea Devil* had the tightest crew I had ever seen. We continually backed each other up on the boat, and we were a close-knit group in port. We also had formed a formidable tackle (without pads) football team that kicked everyone's stern cheeks on weekends in port.

The *Sea Devil* also had one of the best leadership teams I had ever seen. CDR Rich Mies (the CO, or commanding officer) and LCDR Dave Cocolin (the XO, or executive officer) were outstanding leaders: firm but fair. Although they demanded high standards and possessed steely-eyed killers-of-the-deep attitudes, their words and actions conveyed that the crew was paramount. Both men went on to have successful careers: Mies retired as a four-star admiral, and Cocolin retired as a captain. The navigators on board, LCDR John Bird and LCDR Bill French, were equally exceptional; both retired as vice admirals. Our leaders had high standards they clearly communicated, and their work ethic was impeccable—I don't think they ever slept. LTJG Jamie Foggo, a red-haired lanky Sailor, was typically assigned as the officer of the deck during my quartermaster watches. We made a darn good navigational team in an often highly stressful environment. Foggo is currently serving as a four-star fleet commander.

We had a young junior quartermaster division on the *Sea Devil*, led by a first class petty officer who wasn't used to the swift pace of the fast-attack life and a chief who wasn't fully engaged. In the middle of a Mediterranean deployment, our QM1 was relieved, and QM2 Rick Rose and I were charged with running the division and figuring out how to best support the mission. They called us "the QM2 Ricks," and we were both cocky young quartermasters who continually tested authority but delivered results. We weren't completely out of line, but we were certainly headed down that path.

When I came on board the *Sea Devil*, I considered myself a one- or two-term sailor. Because of a couple of key events during that time, however, I finished that tour as a career sailor.

The first event was when LCDR Bird asked to see me in his stateroom. He told me I had talent, but I was never going to excel if I didn't drop my crappy attitude. I'd had a few similar lectures in my career by this point, but this was the first time it wasn't just a "butt-chewing"—instead, it was a conversation on the expectations of the command and how I could do better. He told me I needed to change my attitude and move forward as part of a team. A few months after that visit with LCDR Bird, I ended up receiving the Sailor of the Quarter award, presented on the ice after we surfaced at the geographic North Pole. When I became MCPON, I told then-Admiral Bird that it was his fault.

The other key event occurred as I was ending my time on board the *Sea Devil*. I had hurt my lower back, and the doc was only giving me Motrin. I was in so much pain that I would lie in my rack many nights and cry, but Doc said he couldn't get me in to the hospital for treatment. A few days later, the CO called me in to his office to talk about re-enlisting. I was in so much pain, I just looked at him and said no.

"I'm not staying in, Sir," I told him resolutely. "My back hurts bad, and no one gives a crap."

CDR Mies picked up the phone in his stateroom right then and there, and the next day I was sent to Charleston Naval Hospital. Just calling the hospital immediately, with me sitting right there, was simply outstanding—it showed me that the commanding officer cared for his crew.

On another occasion, the captain again showed how a good leader takes care of his team, even when that means making an unpopular decision for some. It turned out that many of the sections hadn't had enough rest and time with their divisions—they were stuck in long periods of duty on port-and-starboard watches, because not enough chiefs were aggressively pursuing qualifications in their traditional watch stations. As a consequence, CDR Mies secured liberty for the Chiefs Mess[14] during a port call. News of the CO's order quickly made its way through the rumor mill and boosted the morale of the rest of the crew. It also provided a wake-up call to the chagrined chiefs to pursue qualifications; from that day forward, and with a few personnel changes, the Chiefs Mess became engaged.

I saw quite a few other instances of great leadership on the *Sea Devil*, and their attitude and my experience on that boat is why I stayed in the Navy. I took those traits with me and tried to pass them down at every opportunity. Even though we never won the Battle "E,"[15] we had a great group of warriors who worked well together. We often joked that we were the best "average" boat on the Charleston waterfront, but we always got the mission done.

* * * * *

In 1988, I was selected to be a chief petty officer (CPO) while on my first shore duty in London, with less than eight years in the service. This was another time I was out of my comfort zone, but I consider the day of my pinning as one of my greatest military accomplishments. Navy chief is the best damn job I ever had—but it's also the hardest, if you do it right. If you ever get to a point as a CPO where you feel all is well, your division

---

[14] "The Chiefs Mess" refers to the berthing and eating space for CPOs, but the term is also used to mean the CPOs who reside there. "The wardroom" has the same connotation for officers: it's the cabin or compartment for naval commissioned officers, but also refers to the officers who occupy the wardroom.
[15] The Battle Efficiency Award (commonly known as the Battle "E"), now the Battle Effectiveness Award, is given to U.S. Navy ships, submarines, and other units who demonstrate overall readiness of the command to carry out its assigned wartime tasks.

is clicking along, and you don't need to make adjustments, then it's time for you to either transfer or retire. You can *always* make positive changes to make yourself and your team better.

The responsibility placed upon a Navy chief is the same whether you have seven or 16 years in the Navy—it's what you do with it that makes the difference. During my time as a chief, I read the Chief Petty Officer Creed and Navy Core Values often, and still do even in retirement. It keeps me grounded to the principles of leadership and being a humble leader. I loved the leadership, the challenge, and the motivation of wearing fouled anchors[16] on my collar. For 25 years, I felt that honor every time I placed the combination cover on my head. Those who came before me and sacrificed their time to make me better fueled my motivation to return that example ten-fold to the Sailors under my charge. I can't say I always got it right, but it wasn't for lack of trying. Success as a chief is not about the awards or accomplishments you wear as part of your uniform—if you're doing it for the prestige, you're in the wrong game. The success that a Navy chief enjoys is gauged by and reflected in the success of those under your charge. It's not about me—it's about *we*.

"Anchor up" is an idea I wholeheartedly support. For the Chiefs Mess, the phrase refers to how CPOs step up to ensure the command operates smoothly. Our Chiefs Messes are something to behold when they are clicking. It's a special group that will always rise to the occasion to support the Sailors, their peers, and the officers appointed over them. Our CPO Messes are at their best when there is banter back and forth and even heated conversations—both are evidence of camaraderie and passion. Our chiefs are strongest when the COB or CMC leadership is engaged. Today, the CPO Mess collectively is strong and deeply rooted across the country and the world. I could be anywhere in or out of the country, and if I needed assistance with anything, I could post it on social media. In a matter of minutes, I would have someone step up to provide the support or point me in the right direction.

When I was MCPON, I posted something on Facebook once, and a young sailor responded back with "Thanks, Chief." Many well-meaning folks jumped on the poor kid's case about not addressing me as MCPON and being more respectful. I was not offended in the slightest, though—on the contrary. I will always be a Navy chief petty officer first.

I was very fortunate to work for two outstanding CNOs. Both Adm. Gary Roughead and Adm. Jonathan Greenert are exceptional leaders who guided the Navy through some very rough seas in our naval history. Not only are they superb naval officers, they both agree on the importance of chief petty officer leadership on the deckplate[17] and their role in the success of a command. If you asked them today who their first chief was, they could tell you. They will also talk about their initial meeting with "their" chief and how that chief took them on board, put their arm around them, and said, "Come on, ensign, let me show you how to be a good officer." I have the utmost respect for these two leaders as they continue to give back to the world's greatest Navy.

---

[16] A "fouled anchor" is an anchor that has become hooked on some impediment or has its cable entwined with another object. A fouled anchor symbolizes the trials and tribulations that every chief petty officer must endure on a daily basis.

[17] "Deckplate leadership" is the modern-day term used by the Navy for the type of engaged leader who sets the tone, knows the mission, knows their people, and develops their people beyond their own expectations as a team and as individuals.

For me, the heartbeat and momentum of the submarine started in the Chiefs Quarters. For the boats that did well, the chiefs were leading the charge in just about every way. Without exception, if the chiefs are active and engaged, a command will run like a well-oiled machine. Rear Adm. John Padgett, then Submarine Force commander, once said, "Officers run the Navy, but it's the chiefs who make the Navy run."

*　　*　　*　　*　　*

Many times, the ability to motivate will win the day when it comes to leadership. Motivation by itself, however, isn't enough—I've seen some excellent motivators who could rally a group and foster teamwork, but that didn't automatically make them great leaders.

I love the word HOOYAH—a Navy cheer that builds morale and motivates Sailors, like the OORAH of the Marine Corps and the HOOAH of the Army. When I was a kid, I believed in the HOOYAH attitude, even though I didn't know then what it was called. I was always proud of the teams I was on and excited about being there. That enthusiastic attitude continued in the Navy as I came up through the ranks, and I tried to instill that pride and morale in my Sailors.

As a TRE instructor, I had the opportunity to screen young submariners who wanted to be submarine divers. I visited the U.S. Navy Dive School on Ford Island, Oahu, and saw our young wanna-be submarine divers in action. It was truly an eye-opening experience. Dive School instructors were sharp in their appearance, professional in their actions, and had a positive can-do spirit—all punctuated by their frequent use of HOOYAH! Their enthusiasm and pride made Dive School a positive experience for everyone.

I wanted to create that same optimistic, can-do, motivated environment when I became COB on board the San Diego-based fast-attack submarine USS *Portsmouth* (SSN 707). You wouldn't think a young, energetic senior chief petty officer in the world's greatest Navy would be nervous about a new assignment, but I was. I was lucky—I walked on board a submarine that had a solid crew. I can't say that morale was the best—they were gearing up for deployment and had been hammering hard—but most of them were smart, hard-working Sailors, and they were proud of their boat, "the Mighty P."

Gearing up for deployment meant we went underway a lot to hone our skills, doing all kinds of operations in the San Diego area. But with each underway came an opportunity to build morale, esprit de corps, and a well-trained and motivated group of warriors. It starts at the brow![18] After we tossed the lines off the submarine to the pier, I would gather all the line-handlers at the hatch just prior to going down, and we would get some motivation done.

"We're the best damn submarine on the waterfront, with the best crew!" I would shout enthusiastically. (After all, you are what you think.) "We're going to kick stern cheeks! One! Two! Three!"

Then, all together, we would yell, "HOOYAH!"

At first it wasn't the loudest and proudest, but after a few underways and a few HOOYAHs on the waterfront, I noticed momentum. The crew got a little louder each time. I wasn't too sold that everyone had bought in to the team spirit, so I began second-guessing myself, wondering whether the battle cry was good for morale or if I was hurting it because

---

[18] The gangway; the temporary bridge connecting the ship's quarterdeck to the pier.

the crew thought it was dumb.

One day, we weren't razor sharp passing the lines to the piers—the crew was slow and messy. When coming in or out of port, the Sailors should get their preparations completed and then, looking good in a sharp single file, stand either at attention or at parade rest. But this day, the lads were slow and it was ugly. This meant I wasn't the happiest camper that day, and after the line evolution I told the guys to go down the hatch. But they all stopped and looked at me like something was missing. One of the Sailors spoke up.

"COB, we forgot to do the HOOYAH," he said.

At that point I knew they had bought in. We stopped and conducted probably the loudest HOOYAH in Point Loma history. It reverberated across the hills, and we saw other submariners who were in port stop what they were doing and look our way. It was a proud COB moment for me, and I had many.

One of the sayings we had was "Proud to be Portsmouth!" We took pride in everything about our boat, even the hull number. To this day, whenever anyone in my family sees the time "7:07" on the clock, we loudly and proudly exclaim, "Portsmouth time!"

* * * * *

The Navy's mission hasn't changed that much from the beginning, over 242 years ago. We still deploy on a routine basis, keeping freedom of navigation on the world's waterways at the forefront while being America's "away team," and I was honored to see it firsthand.

I travelled the world for most of my four years as MCPON, seeing what the Sailors were doing and finding out what they needed to be successful. Both CNOs I served with had high expectations and made it clear I was their unfiltered "eyes and ears" as I visited with thousands of our Sailors. I assumed the duties as MCPON on a Friday afternoon; that Sunday, CNO Roughead and I were inbound to Iraq and Afghanistan to visit with our Sailors in the Fifth Fleet AOR.[19] The Navy typically has Seabees, Corpsmen, SEALs, and a few other ratings on the ground supporting war efforts, but in both Iraq and Afghanistan we provide an "all hands on deck" presence to provide key ground support roles through what we called "individual augmentees" (IAs). IAs were pulled from across the fleet (afloat and ashore), and our Sailors performed brilliantly. After every trip, I came away impressed with how hard our Sailors worked as a team, how innovative and resilient they were, and with their remarkable ability to accomplish any mission, anytime, anywhere.

Simply put, our outstanding Sailors are some of the finest individuals you'll ever meet. To the Sailors on watch on land, above, on, or below the world's seas, thank you for what you are doing. You belong to the greatest Navy in the history of the world and you make it better every day. To those who laid the foundation for today's and future Sailors, thank you. I'm equally proud of each of you and glad I can call myself a Sailor alongside you. A large part of any success I've enjoyed is due to the leaders who took the time to show me the way, the Sailors I served alongside or had the honor to lead, and all those—especially my family—who allowed me and our Navy Warriors the opportunity to sail over the horizon and focus on the job in front of us. I'm damn proud to call you Shipmates. Go Navy! ■

[19] The Fifth Fleet is responsible for naval forces in the Arabian Gulf, Red Sea, Arabian Sea, and parts of the Indian Ocean. It shares a commander and headquarters with U.S. Naval Forces Central Command (NAVCENT) in Bahrain.

Lesson Two: Attention to Detail

Good leaders delegate and empower others liberally,
but they pay attention to details, every day.
~ Gen. Colin Powell

I started with some wisdom from my dad. Now I'd like to pass along a few lessons that resonated with me as I made my way up the ranks as a young Sailor and chief. These lessons aren't all mine; I got many of them through the blood, sweat, and tears of those who came before me. When I see practices I like, I just build upon them and pass them down.

1. Get the little things right

Attention to detail is imperative for all Sailors, but particularly chiefs. Leaders need to stay on top of correcting minor discrepancies and should know what Sailors are doing on a daily basis. Little things eventually add up and become bigger and bigger things. Many little issues or items that aren't right will always add up—I used to say "three little things equal one major hit on an inspection."

I once saw a commanding officer get a buzz saw from a three-star admiral during a Submarine Force visit, in front of many of us in the control room. I was a young Sailor, and I didn't like the interaction. I was embarrassed for the CO. The admiral had toured the boat and was upset about screws and bolts: just about everywhere you looked on decking and equipment panels, a screw or bolt was missing. It was just something we lived with, as the equipment worked fine—you might have the two corner screws in place, but the other two out for ease of access and maintenance.

But if the screws and bolts were missing, where else were we taking short cuts? It was an old boat, and details like that mattered. In hindsight, the admiral's concern was justified.

Needless to say, we were embarrassed that we had let the skipper down. We took care of those missing screws and bolts, and went on to be the Battle "E" boat the following year.

"Don't just look at it—fix it!" became our motto.

2. Names are important

I've always tried my best to get names right, in both spelling and speaking. Some might think it's just a little thing, but I believe it makes a big difference.

I had been COB for a few months and took pride in doing the watchbills myself. I posted a bill on the board one day, and a Sailor with a tough name to pronounce and spell was standing beside me. After looking at the watchbill, he said, "Thanks, COB." I thought he was joking, as I had given him the mid-watch—the one in the middle of the night.

"What for?" I asked him, surprised. "For giving you that watch?"

"No," he said with a quick smile. "I've been on board for over a year, and since you've been here my name has been spelled correctly."

I walked away from that interaction thinking, *Wow. The little things!*

3. Procedural compliance isn't optional

Procedures and instructions have been written with a lot of blood, sweat, and tears—so follow them.

When I was a young quartermaster chief, we had a major inspection at the end of a patrol. Thinking we were ready, I focused our team on the basics of navigation. We were undermanned—that's not an excuse, but I had three young QMs and we were bringing one up to speed—and I didn't make sure the paperwork or logs were capturing everything according to procedures. The deficiency wasn't putting us in danger, but we weren't keeping the logs as detailed as we should have. Needless to say, we barely scraped by with an "Average" overall in navigation. In logs, we got a blistering "Below Average."

My guys and I all sat down and discussed how we could do better. Those Sailors were amazing, and they responded quickly. During the next cycle, we got "Above Average." Paying attention to the little things makes you sharp! When I was growing up, my mom would say, "Rick, don't forget to dot your 'i's and cross your 't's." Paperwork is boring, tiresome, and sometimes painful, but it's imperative.

4. It starts at the brow and topside

Anytime you walk down the pier at a naval base, the first thing you look at when you approach a boat or ship is the brow banner: a big sign that shows the name, hull number, warfare insignia, and maybe a cool saying or logo. A wise old crusty Sailor by the name of Soupy Campbell taught me the importance of the brow banner and topside. He was COB on the fast-attack submarine USS *Birmingham* (SSN 695). Soupy probably doesn't know—or doesn't want to accept the responsibility—that he played a big part of any success I had as a COB and beyond.

Anytime you saw the *Birmingham*, the brow was hung ship-shape and the topside watch was sharp. You typically hung the brow banner with zip ties around the accommodation ladder of the ship. What Soupy taught me was that the brow banner is a good sign that the ship is taking care of the details. I used that as a gauge as I made it through the CMC ranks, and certainly when I was MCPON. If the brow banner was sloppy, 90 percent of the time the crew wasn't that sharp, either. This meant their CPOs and wardrooms weren't as engaged as they could be—they were missing the "little things," which led to deeper issues.

So check your brow banner and check your quarterdeck, and if they're not squared away, fix them. If the Sailors don't look or act sharp, fix them. If you walk away from or pass by substandard things, you've just lowered the standard as a leader. It's easy to be the good guy, but it's often hard to enforce the required standards. Why is it we often set liberty dress standards in foreign ports? When a Sailor leaves the ship after a hard day's work, they still represent the command where they are stationed, the leadership of that command, and our great Navy.

What starts at the brow needs to continue through the rest of the ship. I've always been a stickler for cleanliness, preservation, and stowage. I learned as a young chief that those items require daily attention or they will quickly get out of hand. Yes, some might say I went overboard as COB of the USS *Portsmouth* and as the CMC of USS *Preble* (DDG 88). Those tasks were always at the forefront, and I'm sure I drove a lot of Sailors crazy with my emphasis on them. But it paid off, because the Chiefs Mess and the crews took pride in their ride, and the ships looked great and performed even better. Those three tasks are crucial and start the process for a ship-shape command. I used to say you can tell how a submarine will perform within 15 minutes of being on board. It's the attitude of the crew, and how sharp they are in uniform standards and actions.

To enforce standards from the very first minute on a ship, I insisted on meeting every newly reporting Sailor when they checked on board the first time. The new Sailor—officer or enlisted—was not allowed below decks from topside without meeting me (or my designated representative) first. I wanted eyes-on contact to ensure their haircut and uniform were squared away and that they knew our high standards before the crew was introduced to them. I can't count how many times I gave five dollars to a young Sailor with instructions to go get a proper haircut. Sometimes I made them come back after their uniform was squared away. Not only did we enforce our ship standards up front but, more importantly, we also gave that Sailor the opportunity to make a good first impression on the crew. Soon the chiefs started ensuring that the Sailors under their charge were ready when they walked on board, which reflected well on an outstanding CPO quarters.

During my time as CMC of *Preble*, I received a great recommendation from one of the chiefs while on deployment: take a picture of a newly reporting Sailor and email it to their family. Letting the folks back home know their Sailor arrived safely took just a few moments, but it reaped so many positive comments from spouses, moms, dads, and other loved ones. It's the little things! ■

Lesson Three: Personal Growth

One can choose to go back toward safety or forward toward growth.
Growth must be chosen again and again; fear must be overcome again and again.
~ Abraham Maslow

You grow when you challenge those who work for you or with you. Most of the success I've had over the years is a direct reflection of those I've interacted with. Don't ever miss an opportunity to learn and grow!

1. Challenge your Sailors

I don't like the term "taking care of Sailors." I think we as leaders can do better than simple caretaking. Yes, you may have to help Sailors just out of boot camp to show them the way, but at some point they need to take care of themselves. Just like you expect your own children to become productive citizens once you've raised them, you want to challenge your crew to become productive Sailors on their own by growing into positions.

I once discussed this topic with a senior admiral about our Sailors today and how they don't want to be "taken care of." In fact, we bet a beer on the reaction from an all-hands call when I asked the thousand or so Sailors in front of us if they wanted us to "take care of them or challenge them." Let's just say I had a cold frothy beverage after that.

Sometimes you need creativity in the challenging. A head-on order doesn't always work; sometimes you must think about it with some ingenuity to get through to someone. When I was COB of the *Portsmouth*, I had a Sailor who was probably one of the smartest young men I've ever met. The problem was that he had a bad attitude, and he sucked the HOOYAH out of the air when he entered a room. He'd probably had a bad leadership experience in the past, and it was hard to connect with him.

In Navy lingo, this young man was a "SNOB": Shortest Nuke On Board among our nuclear-trained warriors. "Nukes" are Sailors in the nuclear propulsion program, and "short" meant that his time in the Navy was coming to an end. He was planning on leaving the Navy—and it showed. He proudly wore a SNOB belt buckle to prove it.

One day I went through the barracks to conduct a "quality of life" inspection, and was surprised to learn this Sailor had a computer setup that would have made Microsoft proud. This was 1996, and computers weren't in every home. I walked back on the submarine with an idea. In the mid-'90s, the COBs had a lot of administrative work, and I had a hard time keeping up with it.

I engaged this young man immediately by complaining to a fellow chief about computer issues when the young Sailor was within earshot. Well, he took the bait and immediately started providing assistance. I had found what made him tick. Sometimes you must find out what interests people to get them engaged at a productive level.

This young man didn't stay in the Navy, but for his remaining time on board, his attitude was at least more positive than it had been. The bottom line is you can't win all the battles, but sometimes you can sway the battles to better support the end result: mission accomplishment.

2. Challenge yourself

One of the best ways to challenge yourself is to take a leap out of your comfort zone. It might leave others scratching their heads, but never be afraid to step outside your own "box" and do something different, even if the idea scares you.

Rising from being a COB to becoming the master chief working for the two-star head of the Navy's submarine force was a significant accomplishment. But I felt I needed to learn more, so I took an assignment that to some might have seemed a step back; in fact, many senior master chiefs on the MCPON's leadership mess frowned at the idea. After more than 20 years in the Navy, I took the assignment to be the CMC of a new guided missile destroyer, USS *Preble*. It was a return to deckplate-level leadership, but in no way was it a step backwards. I was able to see things from a different perspective—that of the Surface Navy at its finest.

I was as nervous as a cat in a roomful of rocking chairs when I reported aboard, but within an hour I knew this destroyer was the place for me. The officers and crew on board the *Preble* were some of the best I've seen, and I would go to war with that group any day. We were lucky enough to have CDR Mike "The Slotman" Slotsky as our CO. He was tough as nails on his junior officers, department heads, and executive officer, held the CPO Mess to high standards, and loved his crew. We often argued whose crew it was, but the training, care, and feeding of those officers and crew were always at the forefront of his actions. When you peeled the layers of the onion back, he had one of the biggest hearts of anyone I know, and all the officers who walked off that ship were some of the most prepared I've seen.

I reported aboard a few months before our maiden deployment. With a new crew, training and qualifications were at the forefront of our deployment. In fact, we trained so hard that everyone on the ship developed muscle memory and could just about perform some evolutions with their eyes closed. On one occasion during the 100+ days in the northern Arabian Gulf, we trained for a helicopter landing and crash. The next day, a helo crashed on deck with several injuries, including a major one. Thanks to our training, strong warfare qualifications, a stellar group of Sailors, and an outstanding corpsman, super-doc Jeff Dell, we suffered no casualties. They saved the life of a young boatswain's mate who had been hit in the head by shrapnel.

I look back on that often and am certain that if we hadn't prepared for such an accident the day before, it could have gone another way. This is why I take training, damage control, and warfare qualifications seriously. These three elements of being a Sailor don't have to be fancy, but they do need to be controlled and effective. Every "routine" evolution is never routine when operating at sea. You have to be brilliant on the basics! We had a tough deployment, but an outstanding crew kicked stern cheeks and rose to the challenge, earning the Battle "E" for the maiden voyage.

From that job, somehow I was selected to be the Pacific Fleet's fleet master chief, which proved to be the steepest learning curve of my career. Going from working for a two-star admiral to a four-star admiral is a significant step on many levels. Both are busy commands, but a four-star's operational schedule is a beast, and their level of engagement is simply eye-watering. Yes, it was a challenge, but I met it by working a little harder and a little smarter, and by relying on the personnel around me, both senior and junior.

Of course, it doesn't always happen that way. During my career, I saw some folks placed in positions of leadership who shouldn't have been. Many weren't able to meet the learning curve at all, or it was way late in their tour when they did, which significantly affected their production and support to the Sailors. This is why I've always been a fan of process when it comes to the proper selection of people for certain billets. Challenge is good, but on the other hand you don't want to put someone in a position that is too much of a stretch for them. It's OK to say no—and that goes for both those giving and those receiving an assignment. Sure, some can rise to the challenge, but those who don't will cause many to suffer the consequences of unready leadership. We were able to check that issue a bit during my time as MCPON by placing experience as a major factor in the selection process for senior command enlisted billets, and by allowing more time to stay in the Navy to grow into those positions.

Many ask what I would do differently and why. Honestly, I don't have a do-over moment. Like everyone, I made mistakes throughout my life and career, and I've been in situations where things didn't go my way. But I always tried to turn them into positive experiences, or at least learn something from them. You can spend your life looking behind you and wondering what might have been, or you can spend your time figuring out how you would like things to be, and work toward putting things in place to make it reality. As leaders, we need to quit dwelling on the past and instead look forward and affect the outcome. If you don't, you'll not only regret it, but you'll find yourself part of the group that's just complaining about the outcome. The bottom line is you can be successful if you apply yourself.

3. Make self-evaluation a priority

Being a young chief with only seven and a half years was a bit overwhelming, but a solid group of chiefs on board USS *Tecumseh* (SSBN 628) greatly helped me mature into a leader. Our Chiefs Mess was very competitive with each other, which made me evaluate the way I conducted myself daily and pushed me to work smarter to keep up with the pace.

This is when I executed what I called the mirror test: I'd wake up, shower, and look in the mirror before starting my day, and I'd do the same thing before ending it that night. You should be having a daily conversation with yourself, asking if you've done everything you can that day to support your crew's mission success. If you find yourself thinking of things you failed to do, you're on the right track. Ask yourself, *What could I have done to better support my Sailors?* If you use these two daily opportunities to be hard on yourself and expect high standards, you're taking the first step toward leading a successful team. They will notice, trust me, and an ounce of example is worth a ton of advice!

It's important to be a humble leader, but don't live in humility. Golda Meir, the prime minister of Israel, once said to a visiting diplomat, "Don't be so humble—you are not that great." Learn from whatever it is that humbled you, then pick yourself up and move forward. I used that kick in the pants to make myself better, but you don't need to approach everything you do with a humbled attitude.

4. Pass it on

I don't believe in having a formal mentoring "program," but I strongly believe everyone should have a mentor and *be* a mentor, whether in on-going sessions or through informal contacts on and off the job. Take every opportunity to sit down with those you lead, your peers, leaders, and family. Have a conversation—talk about everything from the morning happenings to how to improve what you're trying to accomplish. Learn from the life lessons a mentor can provide—your team will grow as a result. With social media and email so readily accessible, you can always reach out to someone for advice or just a conversation about leadership.

My favorite informal setting for mentorship is a game of cribbage. I'd known the game but didn't understand how it was played until I was in the submarine force. It's a fun game, but more importantly, it becomes a vehicle through which you can mentor and be mentored by those around you. It allows people to let their barriers down and connect with others. I often called it a therapy session. Regardless of your level, from admiral down to seaman, you will benefit from these informal sessions with those you lead and work with. I've found that even in the corporate world, many could benefit from a good game of cribbage—or any other game or sport—with their subordinates. Try it, and you'll be surprised what you learn and gain from that interaction. It's also a time to pass along your vision and leadership and reinforce your standards. In my many years as an inspector and riding every submarine in the Pacific, I saw that those who took time for a quick game during or after chow or watch often had some of the best performance from their crews.

It's useful to regularly reflect on those people who influenced you, as well as how and what they learned from you. Another good exercise is to think about five characteristics or traits that shaped you as a person. Write them down and discuss them with your peers and others around you—you might be amazed by just how rich a discussion you'll have. ■

Lesson Four: Effective Communication

*Communicate downward to subordinates with at least the same care
and attention as you communicate upward to superiors.*
~ L. B. Belker

The most successful commands empowered Sailors to complete their jobs. Good leaders give clear direction, then give others the leeway to do what they need to do. Communication is the key for most of the success—clearly written and verbal guidance can set the tone for the entire leadership team and make the crew part of the goal process. When commands and leadership teams get in trouble, the root cause is typically related to lack of communication.

I took many notes about what to do and what not to do when I became a TRE inspector and then a COB. While riding ships, I always took mental notes, wrote things in a book, and borrowed things like policies, watchbills, and ideas that successfully supported the crew. Why re-invent the wheel when you can just redesign it and make it better?

We saw boats that were struggling. In virtually every case, the leadership team was not communicating to the crew, and they weren't setting clear standards or expectations, leaving the crew to flounder. Failure to communicate with the crew also prevented them from functioning as a unit. Once you go down the struggle hole, it's hard to reverse the trend unless there's a major leadership change.

I remember one boat where the commanding officer was a screamer and the COB was not engaged in the day-to-day happenings of the ship. Both the wardroom and the Chiefs Mess had a good amount of talent, but without communications and overall guidance, the crew was left to falter. This resulted in the crew going from a Battle "E" contender to just struggling to get the job done. Attrition was high and leadership was not as involved as they should have been. It took years to overcome that disengaged culture and lack of standards that had been passed down over a period of time. The lesson here is to engage early and often and not be afraid to lead.

How you treat your team—which includes how and when you communicate instructions and criticism—is an accurate measure of how effective you are as a leader. These principles in particular are important:

1. Don't waste others' time
"Son, the early bird gets the worm," my dad would often say. When I was a kid, I don't recall ever being late for anything.

As a chief, I thought it was important to be the first in and the last off. It's not about you—it's about the Sailors you lead. Whatever you do, don't waste their time. As an E-6 and below, I was always frustrated by how much time was wasted in the morning: we had separate meetings with the department, the chief petty officers, and the leading petty officers. Many times, hours would go by without proper and direct communication, while the Sailors just waited to get the word or work for the day.

One of the boats I served on had work lists generated the evening before, which allowed the Sailors to hit the deck running in the morning. I took that idea and used it during my time as a young chief. The division not only met my expectation of generating the list but also exceeded it; they started adding items themselves when they saw something wrong or that needed attention. It worked! By the time I was finished with the morning meetings, my team was well into their day.

I wish I could take credit for this, but I had a great first class petty officer at the time, QM1 Mark Mandile, who executed the plan and motivated the team to move forward. Not surprisingly, he went on to become a COB and master chief prior to his retirement. I didn't like wasting their time and I also gave them the leeway to get the job completed. If they didn't meet my expectations, then we would re-group to discuss and fix it, but they rarely let me down. Wasting your warriors' or employees' time can be one of the biggest culprits in lowering your team's morale, and frankly can be a deciding factor on whether a young Sailor sticks around in our great Navy.

2. Be grounded and approachable

Don't ever think you're better than those around you. You might be surprised by who they are and what they can bring to the table, given the right environment. Leadership is about taking input from others and using that input to achieve something truly remarkable.

I seem to always go back to the submarine force, as most of my career was conducted below the ocean. When you think about what they do, it can boggle your mind. Every day, submariners take a perfectly good piece of machinery and make it sink for months at a time to accomplish an extraordinary mission. You see something similar in nearly every community in the military and in the corporate world: leaders at work, every day, doing something great. Those who are the most successful are considerate of those whom they are tasked to lead, and they foster an environment for success at all levels.

When you start thinking you're better than someone else, that's the beginning of a downward spiral. When you start to discount the value of what those below you in the rank structure have to offer, that may be the time you start to lose valuable input and insight. I've seen quite a few leaders who stop listening to those they lead once they've made it to the top, and I've also seen those who aren't looking for ways to improve either themselves or the warriors they lead. Those types of leaders frustrate me and typically don't last long.

For me, it was important to bust suds in the galley, clean the heads, and do almost every other crummy little job there was to do. I think it's even more important to not forget how to do basic tasks—it brings a reality check back into your leadership style. Rolling up my sleeves and digging in with both hands let my crew see I wasn't afraid of good old-fashioned Sailor work. It was a chance to set the pace and set expectations with my deeds, not my words.

If you are in a leadership position, think about this hard: When was the last time you did the dirty job? What has changed since you did it yourself, and how can you make it better for others? Just because you did it a certain way doesn't mean it must stay the same. Talk to your peers openly about the idea. After doing this, discuss what you've learned. It might amaze you.

Those in leadership positions must realize there is a need to continuously evolve. It might be new technology that makes it easier or it might be a new way of doing something, or it could even be that the standards or culture have changed, so you have to adjust your own attitude to get things right or completed. You can't forget where you came from, and you can't treat people as though they're insignificant, no matter their role.

My dad would always tell me to say hello to the garbage man and street cleaner. He taught me to treat everyone as though their job was the most important one. I took this counsel on board as a Sailor and navy chief petty officer—I made it a point to say hello to the new Sailors and find out who they were and where they were from. They helped me by providing some valuable insight into my role in the command. In retirement, I still make it a point to treat others as I would want to be treated.

When I was the force master chief for the Submarine Force, U.S. Pacific Fleet in Hawai'i, I came in to work one morning and found a second class petty officer sitting outside the building on a bench. Eager to report to her new duty station, she had buzzed the door at 0600. She was told to sit outside until someone came for her. Finally, a guy in shorts, flip flops, and aloha shirt (me) walked up and asked her what she was doing. I invited her to come in while she waited, gave her something to drink, and offered her a seat in my office. We visited for about 30 minutes, then I finally sat down behind my desk. That's when she asked what I did. When I told her I was the Submarine Force master chief, she was quite surprised.

The experience provided great insight and feedback, and we quickly adjusted our sponsor program so Sailors and their families received the oversight and information to start them out successfully. Some of my greatest conversations with and feedback from Sailors occurred while I was in civilian clothes, when they had no idea who I was. My point here is to continually look at all venues to adjust your programs and ways of doing business. Change is good when you control it; don't let it control you.

3. Recognize your people

People need attention. Yes, even us older folks like to hear we've done well—it's human nature.

Qualifying submarines marked my first real success in the Navy. The pride I had when I earned my dolphins has never left me. In fact, I found myself looking at the dolphins on my chest at every opportunity; it was the work ethic and effort I put in, and the support of my team, that allowed me to earn that achievement.

Over the years, the rewards and awards will continue to come if you work hard. Many people measure others by the amount of ribbons on their chest and how many letters of appreciation they've received. Sometimes what means the most to a Sailor is just a pat on the back and a "great job!" The way a leader delivers feedback is what matters—even if it's constructive criticism, you can say it in a way that's beneficial for both parties.

The best time for a leader to visit with their team is typically on the mid-watch, or when a Sailor is working their hardest. That's the time to best see and hear how your people and their families are doing, as well as how leadership is measuring up and getting the job done. The pat on the back goes a long way. Yes, they'll get their accolades for a job done well, but the real memory for them will be the time the boss went down to their level and told them how much he appreciated them.

When I was a young petty officer on a submarine, my CO on the USS *Sea Devil* took the time to do that. CDR Mies certainly connected with the Sailors on board, and still does to this day, long after his retirement. Even after I'd left his command, each time I advanced, I would receive a hand-written letter from him with congratulations. I was amazed he would take the time out of his busy schedule to recognize my efforts and successes. In fact, I've since learned that others who worked their butts off for him also received those letters. To me, the most amazing fact is that he did this as a four-star admiral with many responsibilities. Even after all his success, he still felt that those who brought him along were important enough to recognize.

Signing letters of recognition or promotion is also a powerful tool to give your people attention. Each August when I was MCPON, I sent a hand-signed letter of congratulations

 TWELFTH MASTER CHIEF PETTY OFFICER OF THE NAVY
MCPON (SS/SW) R.D. West (Ret)

Bravo Zulu on your selection to Chief Petty Officer and welcome to the Chief's Mess!

As Chief Petty Officers, mission (warfighting) is our #1 priority, but giving the Sailors under your charge the leadership, motivation, tools, and training, to succeed will ensure mission accomplishment. Without our people, the command is not whole—it has no mission, no ability to operate, and no role in the defense of the nation.

As Chief Petty Officer's your charge, your foremost responsibility, is to your people, "your team", officers, enlisted, and family members assigned. In order to carry out the mission and bring every shipmate home safely, you will be charged with achieving and maintaining the highest levels of operational readiness, training, personnel and material readiness within the command you are assigned. To do this it will require synergistic leadership within the Mess. The following leadership principles will provide a solid foundation for you to be a productive member and support the Chiefs Mess:

- ⚓ **Maintain a High Level of Professional Knowledge**. Know and understand your job, equipment, and the capability of your personnel. Never miss an opportunity to train.
- ⚓ **Conduct/Utilize Procedural Compliance**. There are formal tactics, techniques and procedures for just about every evolution our Navy conducts. Know them and follow them exactingly!
- ⚓ **Value Integrity**. You must have the moral fortitude to do the right thing at all times, don't take any action on or off the ship that may dishonor yourself, the command, the U.S. Navy or our great Nation. If you do make a mistake, admit it, study it and learn from it, just don't make a habit of it.
- ⚓ **Provide Forceful Backup**. No one person is infallible. When a Shipmate is steering the wrong course or is about to, stand-in and provide forceful back up! As a Chief this applies at all levels of the Chain of Command to include CO, XO and CMC.
- ⚓ **Have a Questioning Attitude**. If something does not look right or sound right, if probably isn't right. Do not accept things that are not, or doesn't appear correct. Never walk by a problem without taking action or you've just re-set the standard for those Sailors around you. Set your high standards early and enforce them often. It starts at the brow!!
- ⚓ **Demand Formal Communications**. There is no routine day in our Navy. Every evolution/event or order must be communicated in a clear and concise manner. Trust but verify! Don't lead from your desk, engage your Sailors by being up-front and center.
- ⚓ **Take Ownership**. Maintain your division, department and command at the highest state of overall readiness. If something is broken, fix it. You ARE a leader at all times.
- ⚓ **Challenge your Warriors**. Qualifications (shipboard and watchstation), and training should be a priority and should be conducted and reinforced daily. You have to be able to "fight the ship".

 Anchor Up Shipmate,

R.D. West (MCPON #12)

"The Navy has both a tradition and a future—and we look with pride and confidence in both directions"
Admiral Arleigh Burke (CNO, 1961)

to every newly selected chief petty officer when the names were announced. The number of letters was in the thousands, but I wanted to sign each one by hand to recognize the new chiefs. No, I didn't use autopen, and yes, it was time-consuming. The staff spent a couple of weeks generating these letters and sent me home with stacks each night; I would sign them late at night, early morning, and during lunch. Receiving a hand-signed letter of promotion meant a lot to me when I was a young Sailor, so it was important to do the same for those advancing through the ranks behind me.

4. Embrace technology and use social media properly
When I joined the Navy in 1981, we communicated by face-to-face conversations, regular mail, telephone calls, and naval message. When underway on a submarine early in my career, we typically only received the news that leadership wanted us to know. For family communications, we relied on something called "family grams," one-way messages about 30 to 40 words from the family or spouse to let the Sailor know all was well back home. These messages were quick and to the point and could not contain anything negative; they were screened by the squadron or group commands before putting them on the formal communications wire for transmission to the boat. Families could only send about five over the course of a three- to six-month deployment.

In 1984, I remember a young DS3[20] on the *Sea Devil* who decided to bring his newly invented "laptop" onboard for deployment. It took about three of us to bring it through the hatch, and we all had a laugh about why he would even buy such a thing. In the mid-'90s, many COBs used pencils for their watchbills and other administrative work, but when I became COB it was common to be assigned a laptop for on-board use. Now just about everyone in the Navy has a personal mobile device that has the computing power far greater than many of the large systems we initially had.

My point is, don't be afraid to embrace technology. In fact, CPOs and officers should be leading the charge. When I took the seat as MCPON, social media had not been widely used in the armed forces. My awesome staff and I started using social media to communicate across the Navy on important topics and to interact with the Sailors. This was one heck of a challenge. My first and biggest lesson was that it was impossible to answer everyone, but I sure tried—it was not uncommon for me to be in my office at 0500 or still up at 2200, at home or on travel, answering questions or concerns. I was lucky I had a great staff that kept me up to speed and supported the pace. I found that junior Sailors, especially, needed and welcomed this kind of communication.

Today, social media is prevalent throughout all ranks and used for almost every occasion to get the message out. It can be an extraordinary tool. When our personnel command in Tennessee flooded, for example, the staff and family support groups used social media to provide real-time updates to support the families and recovery efforts. Social media can introduce a Sailor to his new baby and help families stay connected during deployment.

On the other hand, social media can be dangerous. Here are some problems to watch out for:

---

20 Data systems technician third class

• There's an old saying: "Loose lips sink ships." Make sure that everyone you lead is careful about showing the world our ship movements, operating environment, and discussions about arrival and departure. Operational security (OPSEC) must always remain at the forefront. A lot of enemies troll social media, looking for personal information about our military and the operations they are conducting.

• When you make yourself public via social media, you are exposed and often subject to criticism. I enjoy feedback, and I welcomed the interaction with identified Sailors who provided constructive, clear, unemotional criticism.

• Be responsible and courteous, and always be mindful that things you post could be used against you in the future. As soon as you hit the "send" button, you are out there! Stand by!

• Social media is a great venue for immediate feedback from families and Sailors. Many leaders thought it would circumvent the chain of command, but my staff or I would provide the communication back through the command or would advise the Sailor to speak to their chief.

• Don't believe everything you read. Check with a credible source before you ramp up your emotions on a particular subject.

• Nothing—*nothing*—compares with face-to-face communication. I've seen good leaders fail because they tried to lead from the computer and didn't interact with their warriors in person. There is nothing like a discussion with a Sailor on a mid-watch in a lower-level engine room on deployment to really gauge how you're doing as a leader. ■

Lesson Five: Attitude Is Everything

The greatest discovery of my generation is that a human being
can alter his life by altering his attitudes of mind.
~ William James

Opportunities may come and go, but in the end, life is just what you make of it.
You will fail, and it will happen more than once. You'll also be judged by your
failures and it will sting, suck, and hurt. As MCPON, I saw plenty of what I
call "ankle biters": people who have nothing better to do than to judge or second-guess
others, based on little to no data. You'll have to learn to deal with these types of people,
especially the numerous trolls on social media. Go to the source, or *be* the source, of solid
information before you cast blame or add to the rumors. Don't let the ankle biters guide
your thoughts.

I've often heard these types lamenting that chiefs have lost their power. Frankly,
that's BS. Chiefs will always have the power to do what's best for the command and for
their Sailors. If you feel your power or authority has been diminished, that means you're
not working hard enough to execute the leadership you were given, and at some point
the power eroded. If that has happened to you, it's certainly not the end of the world—get
over yourself and get back in the game by engaging the Chiefs Mess, the wardroom, and
command leadership. Chiefs really do have the opportunity to turn a command into a
positive experience through their leadership and actions. If it's in your wheelhouse and you
feel you can make a difference, then you must not only control it but own it, or spend the
rest of your life wishing you had.

Like many Sailors, I almost got out of the Navy after my first tour. I got caught up
in the moment and influenced by some of the naysayers about how things suck from day
to day. I look back on that now and can barely remember the bad. But at the time, I let
the negativity of the few affect me. Since then, I make an effort to avoid negative people—
they can bring you down or bring down those on your team. Instead, I try to surround
myself with positive people, those who want to continually make improvements and seek
opportunities.

An upbeat, glass-half-full attitude will have a positive impact on those around you.
The best part is that it's contagious—those around you will develop a better attitude and
pass that on to others around them.

Whether you're the Sailor washing dishes on board a nuclear-powered aircraft carrier
or the captain of that ship, your job is important and your actions and attitude should
reflect that. I guess it was the way I was raised—from my own humble beginnings, I've
always thought that the lowest-ranking person in any command or organization is just as
important to team success as the boss. Their input and enthusiasm can certainly bolster the
team effort, while the lack of a positive attitude can drag it down.

One of the many great experiences I had while serving as MCPON was flying out to
an aircraft carrier in the Arabian Gulf that was conducting strikes against the enemy. The

ship was a flurry of activity as ADM Greenert and I made our way around the ship. We usually didn't make rounds together; our mode of operation was almost always "divide and conquer." I stopped through the ship's galley where the food was being prepared for the crew. It was a hot day, 120-plus degrees on the flight deck. Despite the air conditioning, it was almost equally hot below.

I found a young Sailor who had a positive way about him. I stood there watching him wash dishes, and you could just tell by the way he worked that this young man would be successful. I stuck my head into the area, and it was steamy hot from all the activity. This Sailor was doing one of the crappiest tasks in the Navy—I know, because I did it for many days early in my career.

"How you doing, Sailor?" I said.

He looked over at me and flashed a big smile.

"Hello, MCPON! How are you this great day?" he asked.

I was expecting something totally different, so he caught me off guard.

"What's going on in there, Sailor?" I replied.

His response again surprised me.

"It's a glorious day, MCPON. I'm launching aircraft!"

I was thinking about a comeback, when it dawned on me what this Sailor was talking about. I asked him to explain what he meant. It was obvious he was washing dishes and not launching aircraft. He said his chain of command had told him that his role was a big part of what they are doing. If he didn't do his job, it probably wouldn't have stopped the jets from flying, but without him pulling his weight, the team couldn't fully function. He had been taught that anything he did would affect mission accomplishment.

The command had clearly communicated their mission while ensuring that each member of the team knew exactly what was expected of them to support the mission. In the Sailor's eyes, his role was just as important as the captain's, even though his responsibility was much smaller. Both the CNO and I left this ship amazed at just how informed and motivated the crew was. Keeping the crew up to date and engaged made the difference between a team and just a group of individuals. It didn't surprise us that the ship earned the Battle "E" for the best in their class that year.

By laying out their expectations and explaining what was going on, leadership of that ship positively affected the morale and, most importantly, the attitude of individual Sailors like the one I met. You see how the little things add up?

One of the most frustrating things for both a young Sailor and a seasoned leader is to hear "I'm not surprised that happened" or "I could have seen that coming." Come on, man! Those words imply you could have done something but didn't. Trust me, I've been as guilty in this area as anyone. You just have to condition yourself so that when you register those thoughts, you think deeper and then put things or actions in place to make corrections early in the process.

If you know about an issue on Monday but wait until later to act, then it will become a major problem by Friday. By then, things have spun out of control, and people who could be there for you are not readily available. You're paddling up that creek alone, and you hope that the collateral damage will be minimal on Monday. From the day I joined the Navy through my time as MCPON and even today, I hear stories about bad things happening to good people because engagement was lacking or non-existent. If someone had acted early, perhaps they could have saved a career and prevented hurting so many others.

Early in my tour as MCPON, I started keeping statistics on CPO leaders who got in trouble. After a few months, I realized we weren't firing most of these chiefs because of work performance, but for personal misconduct. More often than not, it was stupid little things—and they clearly knew better.

Your position of authority stops when you head down the knucklehead trail. When I looked at the number of chiefs who ran aground, I wondered what would have happened if someone had only given them a friendly reminder or suggested a course change when they were steering into shallow water. Nothing is worse than seeing folks work so hard to get where they are and then throw it away by doing something just plain dumb—but I'm even more stunned by those who knew their shipmates were steering into danger but didn't do anything to get them right and back on track.

It's easy to be a bystander and watch the train wreck. I prefer to be a do-stander and try to affect what's happening around me. A do-stander is someone who engages early and often with something or someone they know will eventually affect them. If a chief, for example, sees another chief making a mistake, they should figure out early how to engage—and then do it! It's not always easy, and sometimes it's the most difficult path, but it's the right one.

You don't have to be a jerk when you suggest a course correction. The one making the error might appreciate the backup—or will, later—and you might just save their career, while also saving the face of the CPO Mess or command. Even in the civilian world, you have an obligation to do what's best for the employees and the company. Of course, if you engage and they tell you to mind your own business, then you have to decide whether to take that intervention further. But at the end of the day, you've done the right thing. Get involved and stay involved! ∎

Lesson Six: Rely on Family

*Families are the compass that guide us. They are the inspiration to reach great heights,
and our comfort when we occasionally falter.*
~ Brad Henry

The ones who have had the most influence on my career and are the keys to any success I've had are my wife, Bobbi, and my sons Zach, a Navy diver, and Nick, a systems analyst for a Navy contractor. I could not have accomplished what I did if it hadn't been for them and their support. Bobbi is the foundation of our family: she keeps our sons and me grounded and focused. When I was gone, which was often, she provided the right amount of discipline and guidance needed for our sons to thrive. Bobbi was also physically and emotionally tough—while I was MCPON, she volunteered and deployed to Afghanistan, spending four months at Camp Leatherneck in support of her administrative role with NCIS.

There were times in my career when I was feeling down or out, and she and the boys would immediately be there to pick me up, sometimes telling me to get off my butt or just get back down to earth. She also did not—and still does not—let me get too big for my britches. The teamwork, hard love, and support we have for each other can't be measured.

We recently celebrated 30 years of marriage. We met in the Navy when she was a Seabee and I was a Submariner—yes, an odd combination, but you know the saying. I would do it all again the same way, with absolutely no regrets. We had many great moments in our young marriage and we're still making them. I'm a lucky guy—we make a great team, and she's still my girl!

Our sons are great young men, and we couldn't be prouder of what they are accomplishing. I've told them several times that if I had one regret, it was all the times I was gone. When I try to tell them, though, each time they cut me short.

"You were always there with us when we needed you, so stop feeling sorry for yourself," they'll say. "Besides, we turned out OK."

So, to all the Navy spouses and families out there: keep pressing on. Yes, it does get tough, but there is a huge amount of strength within your family, those alongside you, and those who went before you. The Navy family is a worldwide network that can and will solve any issue together.

One of my fondest but hardest memories as a young married chief was while I was stationed on board USS *Tecumseh*, homeported in South Carolina. The boat would come into Charleston Harbor for a turn-around to bring on personnel, supplies, or other things. At times we had been out for months; sometimes we had only been out for a week or two, but were getting ready to disappear for a few months.

But as sure as the tide, as we entered port, Bobbi would bring our two sons to a part of the beach where the boat loitered offshore while we conducted the turn-around. I was the periscope operator as a quartermaster on those occasions, and in between rounds taking bearings on navigation aids, I would take a quick glance and see them non-stop waving from the beach. They certainly couldn't see me but knew I would take a few seconds for a glimpse. They would stand there until we turned the bend to head out to sea.

I don't think my men knew what was going on, but they might have wondered why the chief's eyes would well up with tears just from the strain of being on the scope. I don't think I ever told Bobbi how much that meant to me, and I know my sons really didn't comprehend what they were there for. But they were there, and that memory will always be one of my fondest.

When I served as MCPON, I was often asked how I made the family and my marriage work. It's not easy; you and your spouse both have to work at it daily with a little give and take. My advice is simple: work hard, but use every opportunity to engage with and support your family. Family time is vital. Thank you, Bobbi! ∎

Lesson Seven: Transitioning

Change is the handmaiden Nature requires to do her miracles with.
~ Mark Twain

Transitioning to the civilian world has been the steepest learning curve of my life. You go from moving at 100 miles an hour on a road you have come to know to a much slower pace in uncharted territory. It can be both rewarding and humbling. I think the military does a good job of teaching to the next level, and along with the education comes experience that allows you to work in more challenging positions. When I struggle today, I have to step back and remind myself that I've only been at this for a few years. In many ways it's like being a junior Sailor again: not knowing much, but willing to learn.

I have a few tips for those who are either retiring or getting out of the Navy:

1. Fight to stay in the Navy!
Do 30 years if you can. The civilian jobs will always be out there—the private sector thrives on great leaders, especially those with a stellar military background.

2. It's OK to not know everything at first
You will learn it in time—this is what made you successful in the military. Be prepared, however, to be frustrated. You aren't the sole point of reference anymore.

3. It's usually not urgent
Even though you may think it's urgent, remember it's a *different* urgent than it was in the military. Lives are usually not on the line in the private sector.

4. It's OK to change jobs
Don't settle on one and stick with it just because you feel you must. We all had a sense of loyalty from our time in the military. You were under a contract then, but now you have choices. I'm on my third job since retiring, but I really like the companies I've worked for and the jobs I've done.

5. Be prepared
Everyone tells me this, but what the heck does it mean? Even though I landed well, I should have networked more aggressively to see more options long before I got out of the Navy. A lot of cool jobs are available; you just have to be assertive and use every opportunity to network. You will likely land a job through people you know.

6. Be flexible
If you tie yourself down to a specific location, you may limit the quality of job you get.

You might have to decide between a higher-paying job or a location of your choice—or you might be lucky and get both. You've talked quality of life and quality of service your entire career, and now you have a greater say in how you get that.

7. Don't be afraid to challenge your comfort zone
From what I've seen and heard as I talk with veterans, those who are the happiest are doing something totally different from what they did in the Navy. I'm completely out of my comfort zone and my work is challenging, but it's also satisfying.

8. Talk to those who have transitioned before you
Buy them a cup of coffee and have a good sit-down conversation. Those who have gone through the process probably have some great advice to pass along. Have your questions and concerns ready—keep a list.

9. Find a hobby
I didn't do this very well while I was in the Navy, so I'm making up for lost time now. Hobbies add to your quality of life. The Navy offers a robust set of opportunities for you to take on a hobby or get out and see the area you live in—access and use Morale, Welfare and Recreation (MWR) often.

10. Find an advisor to help you stay or get financially healthy
This is one you need to start early in your career. Did you know the Navy Relief Society provides a free budget plan and advice on finances? I've sent hundreds of Sailors to them throughout my career and into retirement.

11. Fine-tune your Internet and social media skills
Get your technology skills up to speed, and make sure your social media game is above the fray. Delete overactive and drama-filled social media pages. Is your email address professional? I once had a guy deliver his resume to me, and I immediately noticed that his email address had several curse words embedded in it. Keep your address simple—just use your name.

12. Stay fit in life and health
There's no excuse for not staying in shape!

*   *   *   *   *

After I retired from the Navy in September 2012, Bobbi and I moved to Poulsbo, Washington. I was lucky and grateful to land at Progeny Systems Corporation, a privately held, high-tech corporation headquartered in Manassas, Virginia, widely regarded for innovative technology and engineering solutions that improve warfighter capability while reducing total ownership cost for the Department of Defense, government agencies, and commercial clients. At Progeny, I work in software and applications technology as a subject-matter expert developing software solutions, and provide leadership and business development activities. Although I work completely outside my Navy training, the work is interesting and enjoyable and is aligned with my Navy passion of supporting and contributing to the warfighters, giving them the edge in a fight.

I'm also employed alongside the five other former senior enlisted advisors of each service as a military advisor for Veterans United Home Loans (VUHL), the leader across the United States in VA home loans. Our focus at Veterans United is educating both active military and veterans on the existence of the VA loan, along with ensuring that the Veterans United staff is aware of service-unique characteristics. Veterans United is the only company I'm aware of that has hired senior enlisted advisors from each of the services.

I remain heavily involved and invested in organizations that support our Navy and veterans. I worked with United Through Reading, an outstanding program, for several years. My current organizational involvement includes the Submarine League (both local and national), Submarine Veterans, Navy League, Naval Studies Board, and the Armed Forces YMCA. One of the most enjoyable groups I have the honor to engage with and support is the Navy Chief Petty Officers. I routinely visit various commands throughout the world, supporting them and giving motivational talks.

Several hobbies keep me busy and engaged within my community. Bobbi and I do woodworking together—we design and make cribbage boards, signs, wine stoppers, and other items, and donate them to various organizations, veterans groups, veterans homes, and to just random people.

I was lucky enough to work alongside five of the most professional warriors during my tenure as MCPON. In 2017, the six of us formed Summit Six LLC. We aim to make this book the first in a series of leadership writings; sharing leadership skills and giving motivational speeches to both military and civilian organizations is at the forefront of our business plan. Not only do I work with these professionals, I have the honor to call them friends, and we are bonded for life through the greatest military in the world. ∎

ABOVE: On board USS *Constitution*, the Navy's oldest commissioned ship, during a rare underway and turnaround in Boston Harbor.

TOP LEFT: Every Thursday at the Submarine School in Groton, Connecticut, leadership would march all the students to the theater on base to discuss liberty over the weekend. I'm not yelling—I'm motivating!

TOP RIGHT: Visiting ND2 West at Mobile Diving and Salvage Unit 2. Our Navy Divers are a special breed of individual. HOOYAH!

ABOVE: In Seattle after my retirement. A funny family moment: while Zach went outside to take a call, we changed into our "Go Navy" attire.

LEFT: My boys, Zach and Nick, testing out the submarine. It's never too early to train them!

TOP: In 2009, four female warriors were our Sailors of the Year; no one noticed that all the SOYs were female until I saw their names on my desk. Proud of these Sailors!

ABOVE: Sailors motivating me on the deckplate. I typically ended all-hands calls with clapping/flying push-ups and a HOOYAH. This day, the decks of the ship were very hot but we did it. Sailors are tough!

RIGHT: Team West at a family gathering in San Diego.

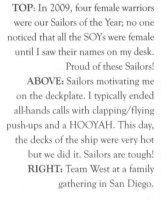

TOP LEFT: I made it a point to hand-sign every CPO selection letter. It's a great accomplishment in a Sailor's career and I wanted to personally highlight it with my signature, not an auto-pen.

TOP RIGHT: In Groton, Connecticut, for Veterans Day. It's important for us to never forget those who laid the foundation and paved the way in our great Navy.

ABOVE LEFT: On the flight line in Kandahar, when Bobbi deployed to Afghanistan with Navy Criminal Investigative Service (NCIS) during my tenure as MCPON. She spent several months deployed in the Helmand Providence and this was the first time I had seen her in four months.

ABOVE RIGHT: With Bobbi at my retirement from the Navy, Sept. 2012.

LEFT: Trust your Sailors. The Sailors from the Navy's Ceremonial Guard in Washington, D.C., conducting the four-man drill performance with me at the center (commander of troops) being very still. I had already shaved that morning.

BOTTOM: A reporter interviews me about Chattanooga Navy week. I love talking about our great Sailors and the best damn Navy in the world.

James A. Roy

J ames A. Roy was appointed 16th Chief Master Sergeant of the Air Force on June 30, 2009, and served until January 24, 2013. He was the first CMSAF selected from a combatant command.

Roy grew up in Monroe, Michigan, and entered the Air Force in September 1982. At stations throughout Florida, South Korea, Missouri, Guam, Mississippi, South Carolina, Virginia, Kuwait, Japan, Hawai'i, and Washington, D.C., Roy worked a variety of civil engineer duties and held numerous leadership positions at every level. He served as a superintendent of a military personnel flight and a mission support group before becoming a command chief master sergeant at the wing, air expeditionary wing, numbered Air Force, and combatant command levels. Before assuming the position of CMSAF, Roy served as senior enlisted leader and advisor to the U.S. Pacific Command Combatant Commander and staff at Camp H. M. Smith, Hawai'i.

As CMSAF, Roy worked with joint and coalition teams and focused on deliberate development and resiliency of Airmen and their families. He led the Air Force through the drawdown of Operation Iraqi Freedom, advocated the next generation of enlisted professional military education (PME), and placed a greater focus on the joint force.

A graduate of the Noncommissioned Officer Preparatory Course at MacDill AFB, Florida, Noncommissioned Officer Academy at Kadena Air Base, Japan, and Senior Noncommissioned Officer Academy at Maxwell AFB, Alabama, Roy has received associate degrees in construction management, construction technology, and applied science. He also received a bachelor's degree in engineering management, summa cum laude, from Park College in Parkville, Missouri, and holds a Master of Science degree in human resources management from Troy State College in Troy, Alabama.

Roy's major awards and decorations include the Distinguished Service Medal, Defense Superior Service Medal, Defense Meritorious Service Medal, Meritorious Service Medal with silver oak leaf cluster, Air Force Commendation Medal with two oak leaf clusters, and Air Force Achievement Medal with silver oak leaf cluster.

In 1993, Roy was the recipient of the John L. Levitow Award, the highest award for enlisted Professional Military Education in the Air Force. He was Senior NCO of the Year in 1996, and received the Ancient Order of Chamorro and a special resolution from Guam's 24th Legislature.

Roy and his wife, Paula, have two sons and live in Summerville, South Carolina. ∎

Introduction: Where I Came From

Until you know where you have been, you don't know where you are going.
~ John F. Kennedy

I was born in Monroe, Michigan, a small town about half an hour south of Detroit. The La-Z-Boy furniture company was founded there in 1927, which seems pretty fitting: Monroe was a quiet town, and not much happened there. I'm the last of six kids, and the youngest by 20 years, so I guess you could say I was a late bloomer. I was certainly a surprise to my mom, who was much older than the parents of other kids my age.

My mother was my rock; I always knew I came first to her. She was the person who made me who I am, and I got my work ethic from her. She grew up in the Depression years, and she worked for one of the car manufacturers during World War II when she was just a teenager, building parts to help the war effort.

I had a humble beginning; Mom and I lived in a school bus for a time, and then in a one-room structure with an outhouse. I remember her working three jobs as a chef while I was growing up. She'd go from one restaurant to another during the day, and then to another one late at night. She was our sole breadwinner, and she never once complained.

I was fortunate to have my grandparents close to us. My mom's father—Grandpa Dutt, we called him—taught me how to fish and chop wood, and I hung out with him all the time. I remember chopping wood with him when he was almost 89. He and my mom were great examples of tireless workers—"You only get what you get by working hard," they'd tell me. So even when I was young, I was expected to help: I worked alongside my mom and grandpa, pushing snow off the roof in the winter and doing yardwork during the summer.

Mom treated me with respect; she gave me complete autonomy, and let me solve problems on my own. One summer, Mom and I moved from our little one-room home into a trailer, and we had a well on our property. Michigan is terribly cold in the winter and water freezes quickly, so I installed lights in the well and insulated the 50-yard line that ran to the house. I was about 10 years old. One winter we had an especially bad storm, and the line froze. We had no water coming in, and once the line freezes, you're done for the winter.

I needed to somehow get water to the house, and I had to figure it out on my own. After some thinking, I hooked up a couple of hoses to the hot water tank and left it running to thaw out the line. The hoses refroze after a bit, but I rolled them up and brought them inside, let them thaw, then repeated the cycle. We finally got water. The next summer I dug the whole line up and put in a new one that wouldn't freeze.

I did all that at the age of 10 because it needed to be done. Nobody told me how to do it; I just did it. I had to mature at a young age because Mom needed my help. Grandpa used to do the same thing: "Hey, Jimmy, I need this snow removed from off the roof. Don't fall through!" That's all the direction I got.

There were always plenty of chores that needed to be done, no matter what season it was, and, of course, there was school on top of that. Recently, when I was getting after one of my sons about doing his homework, he asked what I did after school when I was his age. I thought relating to his situation might inspire him.

"I got off the bus, put my books down, and went out and shoveled snow for a couple of hours," I told him. "Then Mom was after me about doing my homework, so I did my homework."

"Oh man," my son answered. "You walked two miles uphill to school, too, right?"

Well, not exactly, but sometimes it felt like it. At the beginning of tenth grade, my mom became ill, so I moved in with one of my sisters and went to school with my nephews, who were more like brothers. I was also fortunate to live close to my cousin Tony, who was my moral compass. He reinforced what I knew was right and what was wrong, and showed me by his example what it meant to be an honorable, trustworthy man. He influenced my path at a very formative time.

Soon after I started at my new high school, I met a beautiful young lady named Paula. We rode the bus together, and pretty soon we started liking each other and then dating. From then on, it was only Miss Paula for me. I kept telling my friends, "I think we're going to make a go of this!" Even though we were only in tenth grade, we decided soon after we started dating that we were going to get married. We just knew.

Of course, we needed to finish high school first, and I needed to make some money. I got a job as a warehouse stock clerk at Rink's Bargain City, but it didn't pay well enough for us to even get an apartment. I knew I could always pay my union dues and get a job at Ford Motor Company, but I definitely didn't want to do that. My sisters clocked in every day to work on endless wheels or bumpers for Ford, and most of my siblings and their families worked for a car manufacturer of some sort. It just seemed so mundane and monotonous, and besides, the car industry in Michigan was starting to take a nose dive. I kept thinking, *Isn't there something else? What else does life have to offer?*

Miss Paula and I discussed our options. We could stay in Monroe, I could work for Ford, and we could settle down and get married. Or . . . how about joining the military? With that thought, it seemed our world opened up, and our path changed forever.

One of Miss Paula's bosses persuaded us to consider the Air Force.

"You guys seem like a very close couple," he said, "and I can tell you from my experience that the Air Force is a better fit for families."

We were 17, in love, wanting to get married and spend time together. I thought of the Vietnam War that had ended less than a decade earlier, and I didn't want to go off somewhere and leave my sweetheart behind. That convinced me to talk to an Air Force recruiter.

"So what do you want to do?" the recruiter asked.

After thinking for a moment, I replied, "Well, I've always had a strong passion for operating heavy equipment. I'd really like to do that."

The recruiter looked down at his folio.

"We've actually filled all those positions," he said. "But I do have something else in mind for you."

"What's that?" I asked.

"I can make you a munitions systems specialist," he said. I didn't know what that was, but I was sold.

Later, I realized that position wouldn't have been a good fit for me. But the recruiter understood my needs and, as any good recruiter should, responded to them. I just needed a job that would pay the bills so Miss Paula and I could get married.

I took my recruiting paperwork to my mom for her signature, since I was 17. She was adamantly opposed to the idea and didn't want to sign. She had worked during the war and had personal experiences with loved ones going off and never coming back, and the thought of me doing that left her in tears.

"Mom, he's grown up," my sister Pat told her. "You have to go let him do what he wants."

Though she struggled with it, that night my mom reached across the table and signed the papers that changed my life. I believe that was the result of divine intervention. Miss Paula and I married just a short while later, in August 1982. After a short honeymoon in Tennessee, I entered the Air Force on September 13th.

Even though she didn't join the Air Force herself, Miss Paula has been an integral part of my military career since the very beginning.

*   *   *   *   *

I didn't understand it at the time, but my path to becoming a good leader began early because of my humble beginnings. You look at who you are by where you came from. So much of my ability to lead came from the hardship and learning I received in little Monroe and Beulah, Michigan, and from the self-sufficiency and respect my mother gave me. She and my grandpa were my examples, and they were the ones who forged my character and enabled me to lead. As I continued on my path through the ranks to become Chief Master Sergeant of the Air Force, I realized that good leadership at the top means remembering where you came from.

That wisdom has often granted me the ability to connect with my Airmen. Those who ascend in the ranks sometimes forget where they began, and they lose the ability to understand the men and women under their instruction. Airmen come from all walks of life—from wealthy suburbs to poor inner cities. Considering the experiences that built me, I've been able to understand the experiences that built others as well. I know I wouldn't have been able to be as effective a leader had I lost sight of my origins and how they contributed to who I am today.

During my 30 years in the Air Force, I've learned some things about leadership that I hope will be useful to current and future leaders. Following are a few of the more important ones. ■

Lesson One: High Expectations

To lead people, walk behind them.
~ Lao Tzu

B asic training at Lackland Air Force Base in Texas showed me two things: one, there's a whole world outside of Monroe, Michigan. Neither Miss Paula nor I had traveled far beyond our little hometown. And two, I realized the military was a great fit for me, because I've always been one to follow the rules. I had a great training instructor (TI) who showed me what it meant to live up to Air Force standards. I didn't want to let him down; I knew he was relying on me to do the right thing. I knew that if I did what my TI told me to do, I would be a good Airman.

Well, like all good Airmen, I always carried a pen in my pants pocket. The summers in Texas are terribly hot, and one day the pen bled all over my pants. I had this big blue spot covering a large portion of my pants, and I was freaking out. I was frantically trying to clean it, and it dawned on me that what concerned me about this pen stain was not being within standards. A stain might seem like a trivial thing, but it was a huge deal to me as a brand-new Airman, because I didn't want to disappoint my TI. I knew he was counting on me to be an example.

My TI brought out the best in me by believing in my potential and not letting me achieve anything less. Because he expected a lot from me, I rose to the challenge, and I discovered that's an essential part of being a good leader. He didn't stand in front and pull me by the ear—he stood behind me and nudged me, if you will, to do my best.

Basic training in 1982 wasn't like what it is today. The one and only time we went out to shoot our weapon, it was raining, so we didn't get a chance to shoot. It wasn't until my first duty station that I shot a weapon; now, Airmen get them within the first week of basic training. The whole idea of being a "warrior" is instilled in today's Airmen from day one.

*   *   *   *   *

Though I joined the Air Force as a munitions systems specialist, I left basic training with the job I wanted from the start: a heavy equipment operator. *I'm an engineer in the U.S. Air Force!* I kept thinking, awed. I knew the good Lord had been looking over us, because I certainly hadn't had a solid plan for my life with Miss Paula until I joined the service.

At my first duty station, Miss Paula and I learned that the Air Force was now our family. We packed up everything we had in one little bitty pickup truck and headed south to our new home at MacDill Air Force Base in Tampa, Florida—only to find out there was a four-year waiting list for housing on base. I was an Airman Basic, bottom of the totem pole, barely earning enough money to make the drive. Miss Paula and I ended up living downtown in an old, small apartment. I worked a side job, and she got a job, and we made ends meet. We were mature enough to manage our money properly; the hardest part was not using air conditioning in hot and humid Tampa. Each time I got promoted, we moved to a better place, so we moved four times in four years.

My experience as an Airman Basic influenced my life for two reasons. First, I never forgot how it felt to be so financially constrained that I had to live in a little apartment downtown with my bride. Later in my career, I had the opportunity to decide who lives on base, how we process new Airmen, and how we sponsor someone into an installation. I didn't want to see another young Airman face the same struggles I had, so I decided that the more junior enlisted needed to live on base, rather than the senior enlisted. That was the opposite of what I'd seen in the military, but it was the right thing to do.

My first supervisor, Tech. Sgt. Nathan Heard, was the second reason my experience at MacDill influenced my life. He was a staff sergeant at the time, and I respected him tremendously. Not only did he set the standard of how an Airman should behave, he also went out of his way to take care of us. He helped me decide where to live—he knew we couldn't afford much, and he didn't want to see us living outside our means. He steered me away from places that were too expensive or in dangerous neighborhoods. Then he went one step further: we only had one car, so he picked me up every morning and drove me to the base. I don't know many supervisors who would do that for an E-1. Sergeant Heard gave me a foundation on which to build my leadership; I learned from his example that being a leader is more than just what you do at the duty section every day. It's about taking care of your Airmen in every aspect of their lives.

At work, Sergeant Heard made sure his Airmen had a solid technical foundation. He did that by giving us challenging tasks and expecting us to do them. He gave me, a brand-new Airman, the responsibility for operating the grader and crane. Typically you don't operate advanced pieces of equipment like that until you've been in the Air Force for a few years. But he saw something in me, and knew I had the ability and desire to rise to the challenge—even before *I* knew I could.

Sergeant Heard instilled a sense of mission in all of us under his supervision. He told us the aircraft wasn't going to take off unless we first went out there and swept up the flight line and the runway. We knew the pilots relied on us, so we had a sense of purpose and duty. Because we had to complete what we started that day, we had a feeling of mission accomplishment when we went home every night. Of course, when you're a civil engineer, you always have something to do—you'll never be caught up. The complete project might not be done, but we didn't go home until we had finished the task he gave us on that particular day. He expected we would do it, and we didn't want to disappoint him.

I learned how to be a leader by watching Sergeant Heard. He looked at the best in his Airmen, sized up their potential, and then made them stretch to reach it. I learned that being a good leader doesn't mean just guiding people below you, it means challenging them to do things that seem hard. When he told me I had 30 days to do my CDCs,[1] I knew I had exactly 30 days, and he expected them done by then. He held me to that. Tech Sergeant Heard is my hero, and I will always be indebted to him.

*　　*　　*　　*　　*

---

[1] Career development course

Another hero who challenged me in the same way was Chief Master Sgt. Johnny Larry. Several assignments into my career, I was sent to Keesler Air Force Base in Biloxi, Mississippi, where Larry was chief of the 81st Civil Engineer Squadron. I had made senior master sergeant, but there were so many of us E-7s that there wasn't a job for me right away. I asked the chief what he wanted me to do. His answer took me by surprise.

"Sergeant Roy," he said, "this is what we're going to do. You're going to write our annual award package for the squadron."

I had just come from Andersen Air Force Base in Guam, and knew nothing about Keesler or the squadron.

"Chief," I said hesitantly, "if you asked me about our unit in Guam, I could tell you everything. But I don't know what the 81st did."

That didn't deter him at all—he seemed to have no doubts about my ability to manage this huge task.

I was perplexed. *Why is the chief doing this?* I thought. *Come on!* Nevertheless, I accepted the assignment without complaint, as a good Airman does. I went around to each of the sections in the squadron and got the information, then put it all together in a package. Chief Larry was pleased with my work, and a few months later he placed me in one of the hardest sections in the squadron. He put me in charge of facility maintenance, which meant I was responsible for every facility on Keesler AFB—all 326 of them. Someone else was already in that position, but Chief moved him so he could place me there.

I'll always be grateful to the chief for making that bold decision—"Hey, I'm going to move Jim Roy into this position; this'll help his career"—and for putting me outside my comfort zone. I didn't know I was capable of such responsibilities, but Chief Larry knew. He taught me that good leaders are aware of their people's strengths and weaknesses, and they look for opportunities that will push individuals to stretch their abilities.

Not too long after that, while I was still at Keesler, I became a chief master sergeant myself. Gen. John Speigel was my wing commander, and he immediately took me under his tutelage. It was early in my career, and he wanted to make sure I kept broadening my experiences, so he sat me down and asked some direct questions.

"What's the plan, Chief?" he asked. "What do you want to do? What's your career look like?"

"Well, Sir, what I really want to do is be a senior enlisted advisor," I responded.

"OK, I think we can make that happen," General Speigel said.

Oh man, I thought, pleased. *That's a strong go here!*

"But you've got to do this job for me first," he continued. "The base needs a superintendent of the military personnel flight."

I must have looked a little confused.

"HR," he explained.

Now, I had been an engineer my whole career, out in the field every day, digging trenches and fixing air conditioning units. My mind and abilities certainly weren't in personnel. I had a radio (what we called a brick back then) and was on call 24/7, including holidays. I was responsible for every air conditioning unit, toilet, water pipe, you name it, in 326 buildings on the base.

"OK, Sir," I said automatically. "If you think that'll work."

That was a defining moment in my career, because I didn't see myself doing that, but General Speigel did. And he knew what it could do for me in the long term.

From that experience, I learned the concept of deliberate development: making sure you have the right person in the right place at the right time, both for the institution and for the individual's continued development. It takes a sharp leader to know what those under his charge are capable of accomplishing, and then to expect them to rise to the challenge when given a new assignment, even when—maybe *especially* when—they never thought they were capable of doing it.

I was able to publicize the concept of deliberate development when I was in the seat, and we began the enlisted force development panels. We were able to put an enlisted member in legislative affairs and in international affairs. Not just assigned there, but as an actual action officer who goes through the training, sits in the office of a member of Congress, contributes as a valuable member to them, and then brings that back to the Air Force. It is good for their careers, and good for the institution. That kind of deliberate development makes us better every single day.

The flip side is that a leader also needs to be aware of someone who is in over their head in a particular position.

One time, I had a subordinate who just didn't have the ability to fulfill his responsibilities. For that particular position, I needed someone who could think on his feet and quickly come up with creative solutions out of the box. I needed someone who was willing to spend time with me and understand me, so he could anticipate what I was thinking and what I was trying to say. The person in that position wasn't able to do any of that.

A good leader, however, doesn't crush a person who isn't performing. What good is that going to do for the mission? The man still had talent—just not for that position. I found something that was a better fit for his abilities. He was an exceptional writer and gifted speaker, so we put him in public affairs, where he excelled.

*   *   *   *   *

Sometimes it's a peer who can challenge you and get you out of your comfort zone. On my first assignment at Osan Air Base in South Korea, I became close friends with Sgt. John Little, who later became a chief master sergeant. We discovered we had a lot in common, but John's an extrovert—a comedian, life of the party, center of attention—and I'm not. After our assignments in Korea were about over, John wanted to go back to Missouri, where he's from. A special duty assignment opened up for the position of instructor, and John wanted me to take it so we could go together.

"Geez, I don't know about this," I told John. "I'd have to stand in front of people and talk to them."

"Aw, it'll be a piece of cake," he responded. "It'll be a lot of fun!"

"Yeah, for you it would be," I said.

"Hey, this'll be good for you," he said confidently. "This'll be good for your career. I have no doubt you'd be great at it."

John explained why the assignment would be good for me, and convinced me to take a chance and do it. Even though I had serious doubts that I could be comfortable as an instructor, John's belief in me gave me the confidence to go for it. I became an instructor for the 3770th Technical Training Group at Fort Leonard Wood, Missouri, and had about 50 to 60 people in every class. Then I became an instructor supervisor, and the experience helped me understand myself a little better. I went out of my comfort zone, stretched a little bit more than what I was used to, and learned how to communicate a little better—one of those must-have qualities you need to be a good leader.

*   *   *   *   *

In 1996, I attended the Air Force Senior Noncommissioned Officer Academy at Gunter Annex, Maxwell Air Force Base in Montgomery, Alabama. I had taken seven weeks off from my assignment as readiness chief for the 36th Civil Engineer Squadron at Andersen AFB in Guam, but in that short time there had been a change of command. My former squadron commander, Lt. Col. Scott Showers, had left for a new assignment, and a new commander, Lt. Col. Stuart Nelson, had taken his place.

While I was at the Senior NCO Academy, I was getting sporadic emails from the base, and bits and pieces from Miss Paula, who had stayed in Guam. We've got this big mission going on, she wrote, and I'm watching the news, trying to figure out what's going on. Communication in 1996 is not what it is today. I learned that about 7,000 Kurdish refugees were coming from northern Iraq and southern Turkey to Guam. We had to build the base up to prepare for an encampment so they could be vetted and sent back to the mainland.

As soon as I got back to Guam from Alabama, I got a call from Lieutenant Colonel Nelson.

"Hey, Senior Master Sergeant Roy, I want to talk to you for a minute," he said. "Why don't you come up to the office."

"Yes, Sir," I responded. *First day back and I have to go see the commander*, I thought, still exhausted from jet lag. I'd barely had time to see Miss Paula.

I got up to his office, and the commander pulled his chair right up to me.

"Sergeant Roy, I've been hearing good things about you," he said enthusiastically. "This is what we're going to do. I'm moving you from Readiness Flight up to Heavy Repair. I'm putting you in charge of all of it."

I hadn't expected that at all. *This guy doesn't even know me!* I thought, stunned. *He has no idea who I am.* I assumed Lieutenant Colonel Showers, my previous commander, had told him something.

"Yes, Sir!" I answered enthusiastically. "When do you want me down there?"

"Well, I'm thinking today," he responded. He wanted me to run the place, and he gave me responsibility for part of the mission with the Kurds.

Once again, I had a commander who was aware of my potential and took care of me, who had a gut feeling that I could move right into a big job with ease. I never would have thought I could handle such a mission, but Lieutenant Colonel Nelson's confidence reassured and motivated me.

Through all these experiences, I learned that a good leader pays close attention to their people, recognizes their individual talents, challenges them outside their comfort zones, pushes them to do what they didn't realize they could do, and fully expects them to rise to the challenge. ■

Lesson Two: Trust and Autonomy

Trust, but verify.
~ Russian proverb

One of President Reagan's mottos was "trust, but verify." I modeled my leadership style on that principle: I trusted those I led and gave them free rein to do their jobs— up to the point where they showed they weren't responsible. And when I verified that someone was slacking or underperforming, I didn't punish them, I helped them.

The successful leaders I've observed have always followed this principle. I've had a lot of good teachers along the way who have allowed me the autonomy to make decisions on my own, and in turn I granted that same independence to those within my stewardship. I encouraged people to make decisions based on their gut feeling and the information they had, instead of waiting for someone to tell them what to do. If I learned they'd been indecisive or hesitant because they didn't have 100 percent of the information, I would teach them that some circumstances require quick actions based on maybe 30 percent of the information, but they have to make a decision. It may or may not be the right one, but someone in that position needs to rely on what your gut tells you, what your experience, training, and education have taught you, and then take action.

When I was facility manager for the 81st Civil Engineer Squadron at Keesler AFB in Mississippi, I gave the Airmen complete autonomy and didn't micromanage them. I taught them correct principles, and then trusted them to govern themselves... until I checked how they were doing and found some problems. At that point, I had to get a bit more involved.

It gets pretty hot in Mississippi, so obviously the air conditioning units need to be working properly all the time. Unfortunately, at one point we were having a hard time keeping the units functioning. We had trainees going through the schoolhouse without air conditioning, making it very difficult to accomplish any training. We were having all kinds of trouble with other equipment failing, too. I waited a bit to see if my team could troubleshoot by themselves, but pretty soon I had to get involved. I went through every mechanical room on the base and discovered an array of problems. I took the leadership of that particular area over to one of the mechanical rooms and showed them why we were having such difficulties. I helped them understand that they had not followed standard procedures, and explained what was acceptable and what was not. I didn't chew them out; I helped them learn how to do their jobs properly.

In the civilian world, I find myself giving the same type of advice that I gave in the Air Force. In one leadership meeting, we were discussing 401K options when someone suggested we give employees the information and let them make a decision. Others insisted that leadership should decide what to do.

I started to offer my own perspective gained from three decades in the Air Force, but an executive interrupted me.

"Hey, Jim," he said brusquely. "The boss already made a decision."

"Let the chief talk," the CEO told him.

"You know, Sir, we lead adults," I said. "It's going to be their decision at the end of the day, and adults make decisions based on information they have. So give them the information and let them decide. If you withhold anything, they're always going to question what else you're hiding from them. Just lay it all out and trust them."

Ultimately, the CEO followed my advice.

*　　*　　*　　*　　*

I had a lot of autonomy growing up. Mom wasn't always there, so she trusted me to do the right thing in her absence. When she was there, she was in charge, but she was very respectful and treated me like an adult. She never talked down to me, even when I was young.

Miss Paula and I do that with our own boys. We've never talked to them in kiddie talk, but in the same voices we would use with adults. Of course, that is what any parent wants their children to turn into: adults who can make their own decisions without having to be told. A parent's goal is to help their children mature and get out on their own, fully capable of making adult decisions without supervision—then come back and visit occasionally. I keep reminding Miss Paula (and she keeps reminding me) that we've got to give the boys a little leeway, even though they're teenagers. We've got to let them make their own decisions. That's the only way people learn.

Every leader—whether you're the head of the family or the CEO of an organization—needs to treat subordinates with respect and allow them space to figure out their own way to fulfill their responsibilities. If you find at some point that you can't trust them on their own, it's time to get involved and help—but only then. ∎

Lesson Three: Strengthen Your Family

No other success can compensate for failure in the home.
~ J. E. McCullough

As much as I love the Air Force, my life revolves around my family. I love being with them, and I love talking about them. My relationships with my mother and grandfather forged who I am, and my relationships with Miss Paula and our boys are my purpose in life. I could not have achieved the success I've had without them.

Any combat veteran will tell you they couldn't have put themselves in harm's way without two things: good teammates who had their back, and a family at home that was waiting for them.

Miss Paula and I have been married for 35 years, and she's been right by my side the entire way. So has her family, and so have our boys and my siblings, nieces, and nephews. My success as a leader is in large part due to the steadfast support and love from my family, including my mom and grandparents, who I believe are watching over me.

Airmen who worked for me before I had children probably didn't like me so well. It wasn't until I became a father that I better understood that people have lives outside the military—they have families, and most have a spouse and children who rely on them. Miss Paula and I wanted one child, but the good Lord had a sense of humor and gave us twin sons, Caleb and Colby. I always said that if He wanted us to have children, He'd give us the patience and other necessary tools to raise them.

If you don't strengthen your family and spend time with them, your resilience in the workplace—whether it's in the military or in the private sector—is often lowered. Even if you achieve success in your career, it won't mean much in the end if you've lost the love and support of your family.

When I was Chief Master Sergeant of the Air Force—"in the seat," we call it—I spent about 275 days a year on the road. Many of us in the military go years and years traveling like that, because that's what the job requires. Still, you can always find ways to stay close to your family. Miss Paula and I tried to include our children in everything we did, whether it was going to a White House formal event, or a social with VIPs at our house. If we could do it, we did. Everyone knew how important my family was to me. I've been asked multiple questions about that: "You've always brought your family to Air Force events. How did you do that, and how did others receive you?" I think people appreciated my commitment to my boys, and when Caleb and Colby got older, they appreciated it too!

I spent a lot of time with Secretary of the Air Force Michael Donley and his wife, Gail, both of whom I greatly respect. Just after we moved into our new house at Andrews Air Force Base, we held a big event. The house is named after the first Chief Master Sergeant of the Air Force, Paul Airey, and pictures of all the former Chief Master Sergeants of the Air Force line the walls of the foyer. There's a powerful sense of history when you walk into the house. Miss Paula had set up a display of Chief Airey and the other Chief

Master Sergeants of the Air Force.

I remember the Secretary and Gail coming into the house for the first time. Gail often tells the story of how Colby, then about 10 years old, gave her a tour of the house. She loved that! The boys are 18 now, but recently Colby said, "You know, I wish I had paid more attention when you were in the seat."

"Colby, you were ten," I said. "How much attention does a ten-year-old have?"

But they still go to ceremonies with me, and Colby—who's planning on a career in public service—especially loves it.

A lot of times, parents wonder if we had an effect on our kids. Was I a good influence? Did I spend enough time with them? Colby made a profound comment a while back that made me reflect on the brief snatches of time I've always tried to make for the boys.

"Dad, sometimes you don't realize that the little things you do in life have an impact," he said.

It made me think of the time I came home one weekend, after being gone all week on a trip visiting Airmen in Singapore and Japan. As soon as I walked in the door, my football player son, Caleb, excitedly asked if I could play catch with him.

"Son, I'm so dog-tired," I said. "I just got back from a trip halfway around the world. Can I play with you in the morning?"

Of course, he was understanding. *How selfish of me*, I thought later. I'm never home, and all my son asks me to do is play catch. I could have gone outside and thrown the football three or four times and he would have been happy. But instead I said no. *I'll never do that again*, I told myself.

Some time later, before Caleb got his driver's license, I told him if he helped me rebuild my old truck, he could have it. I have a garage and shop right here at our house, because I like to mess with cars and equipment. For an entire month, it was just Caleb and me working on the truck. We changed out the motor and rebuilt all kinds of parts for it— just the two of us. He was going through a very difficult time at that point in his life, and it was an important few weeks that we spent together. I look back at it now, and that time was crucial in making a difference in his life.

About a year and a half later, he came home from school one day, driving his beloved truck. (Miss Paula gets after him to turn the air conditioner on because it's so hot inside the cab. "No way, Mom!" he says. "I gotta hear the truck!") He had stopped at a gas station and ran into an old buddy who had not made good choices with his life. When he came into the shop, I could tell there was something on his mind, and asked what was wrong.

"Dad, I saw so-and-so at the gas station tonight."

"Oh yeah?" I asked casually. "How did that go? Everything OK with him?"

"Well, OK, I guess," he answered.

He suddenly reached for me and gave me a big bear bug.

"Dad, thanks for taking me away from all that," he said.

That was certainly a rewarding parenting moment, but I can't take all the credit. Maybe divine intervention had something to do with it.

After I retired, I wasn't traveling a lot, so we spent more time together. We got a chance to know each other better, and were able to have adult conversations. I didn't know I was having an impact when we worked on that truck; I was just being a good parent, trying to spend time with my boys and doing a hobby we both enjoyed. He realized later that it did influence him to make the right decisions. Now we've started a new project that will take much longer.

About the same time, when the boys were 16, Colby was working on three different presidential campaigns. He was interested in politics, so he wanted to come with me to see a candidate up north.

"That's three hours away," I told him. "But if you want to go, I'll take the day off and we'll drive up there."

You can't do much but talk while you're driving, so Colby was asking me dozens of thoughtful questions about Air Force leadership, what does this mean, who did that, and so forth. We had a great time. The event was interesting, but we left before the candidate had finished. On the way back, we had another great talk.

Wow, he's grown up, I thought. *We're having an adult conversation.*

The next day, Miss Paula asked Colby how it went.

"It was OK," he answered. "But you know, Mom, I'm beginning to think it's not all about that."

"What do you mean?" She asked.

"Well, the best part was just the three-hour drive up there and the three-hour drive back and the conversation I had with Dad."

You can't buy that kind of priceless reaction.

*   *   *   *   *

One of the things I realized when speaking with a lot of Airmen in deployed locations was, number one, they missed their families. They are out there to defend America's interest. But if you talk to any of them at length—whether it's an Airman out in the middle of Iraq or Afghanistan on a PRT,[2] a combat medic down range, or a civil engineer building a new taxiway or ramp—you realize it's really about their family. The national interest is the family, and they saw great value in protecting that. Resiliency for an individual Airman, then, depends in large part on the support of a family.

The hardest thing I've ever had to do is talk to a family member who's just lost a loved one. What do you say? In 2011, 30 American servicemen were killed when a Chinook helicopter crashed in eastern Afghanistan. Their remains were flown to Dover Air Force Base in Delaware, where President Obama talked privately with the families. Air Force Chief of Staff Gen. Norton Schwartz and I were there, along with other military leaders, chaplains, and grief counselors. We split up so we could talk to the families individually, and by the end of our time at Dover we were able to talk to all of them.

Each family was grieving at a different stage. One family would say their Airman died doing what he loved; another would say his grandfather had served. Those were the families you could tell were really close. Then you'd visit with the next family, heartbroken with grief, and often they had some kind of family issue going on, and the weight of dealing with this casualty on top of that was just crushing them. Being able to assign casualty

---

2 Provincial reconstruction team

workers to families for whatever they need is a big strength of the military.

The dwell rates, length of deployments, speed turning of deployments, and the different types of locations our Airmen were deploying to all put stress on Air Force families. General Schwartz and I made it a priority to help them through pre-deployment, deployment, and post-deployment; resiliency for both Airmen and their families depends on the success of all three stages.

Miss Paula championed the Key Spouse program, a valuable tool to support families, especially during separations and other emergency situations. A "key spouse" serves as a link between squadron leadership and families. The program is critical to the success of our Airmen and therefore the success of the Air Force mission, because Airmen who have the support of their families are happier and more productive.

For my wife, it wasn't until a certain point in my career that she felt personally attached to the Air Force mission. We have been in some units where we didn't always feel a part of that—or at least the family didn't. But we've been in others where we still stay in touch with the squadron commander because they made us feel a part of that team. I believe that having committed, involved families is the key to a successful Airman; resiliency, key spouses, and comprehensive Airman fitness all go hand in hand. That is something all military leaders need to carry on. ■

Lesson Four: Care at a Personal Level

Great leaders genuinely care for and love the people they lead more than they love leading itself.
Leadership without love degenerates into self-serving manipulation.
~ Rick Warren

A good leader is personal and accessible; your subordinates should know they can approach you with any subject. They should respect you, but not fear you. They should know you're not going to rip their head off if they make a mistake, but rather you'll listen to them and help them if needed.

Lt. Col. Scott Showers, my commander at Andersen AFB in Guam, taught me how to be a personal leader. He would regularly invite his senior leaders over to his house, just to chat and have a good time with them and their families. He taught me how to take care of people. He was open and friendly in a professional way—that's not fraternization, it's personal leadership. He made each person feel important. When he pulled me aside and said he was moving me to superintendent of the 36th Civil Engineer Squadron, Readiness Flight, it was because he saw something in me that he needed for that position. He had taken the time and interest to get to know me personally and understand my abilities.

Adm. Timothy Keating made you feel like you were the center of the universe when you were with him. For two years, I was his senior enlisted leader and advisor when he was the U.S. Pacific Command combatant commander at Camp H.M. Smith in Hawai'i. We traveled together frequently, and one time we stopped at a little bitty island, part of the American Samoa island chain, to visit one Soldier. He was a double amputee, and Admiral Keating wanted to make sure he was getting the care he needed.

Anytime he saw Miss Paula and the boys, he would always come over and greet them warmly. It wasn't just my family—he did that with the family of every one of us on his staff. He would come over to my sons, who were very young at the time, and humbly thank them.

"Thank you for coming out here, boys," he'd say. "Thank you for letting your dad do the things he does for our nation. Do you realize your dad had a huge impact on this mission?"

That was just how he operated: always humble, grateful, and deeply interested in whomever he was talking to.

Chief Master Sgt. Johnny Larry was another leader who went out of his way to show one of his Airmen that he cared about him personally. When my mom died in December 1987, I was a staff sergeant in the 554th RED HORSE[3] Squadron at Osan Air Base in South Korea. A chief master sergeant was the supervisor of my large section.

"After the funeral," he instructed me, "you need to come back and outprocess."

Larry, a senior master sergeant at the time, thought that was unreasonable and said so.

"He's not coming back," he told my supervisor firmly. "This is the end of December,

---

[3] RED HORSE (rapid engineer deployable heavy operational repair squadron engineer) squadrons are the U.S. Air Force's heavy-construction units. Their capabilities are similar to those of the U.S. Navy Seabees.

and he's got to be home with his wife and family until at least mid-January. By the time he gets back here, he'll only have a month before his special duty assignment at Fort Leonard Wood. Just let him go."

Sergeant Larry is one of my heroes. He taught me that sometimes you just have to do the right thing for one person. He went to bat for me, and the commander finally agreed that I could stay in Michigan until the funeral was over, then proceed to my assignment in Missouri. That taught me an important lesson about standing up for principle, even if your actions only benefit one person.

*　　*　　*　　*　　*

I spent a lot of time, while in the seat, going to PRTs outside the wire. I remember how difficult it was to get to some of these locations—first we'd take a commercial aircraft and then a military aircraft. It was worth it, of course, because we needed to understand what challenges our Airmen faced. If you're going to provide the training for those Airmen to go into harm's way, outside the wire every single day, you better understand it fully. If you don't, how can you advocate? How can you champion their training? You simply can't, so I always found it very valuable to visit the far reaches of where we had Airmen. We may have had two or three Airmen on a base, on an installation that was primarily Soldiers or Marines, so it made a difference to those Airmen. They knew that somebody was interested in them.

Another benefit with smaller groups is that they are more likely to give you honest feedback and complaints. Sometimes people just need to know that someone is listening to them.

On one trip, my team and I spent two days in the middle of nowhere with a handful of Airmen who were conducting a special mission. I was with a lieutenant colonel and a senior NCO, and we were visiting a group of young Airmen just barely out of high school. Some of these young people weren't even shaving yet, but they were influencing the outcome of the war. They were sleeping in huts and living off the land, killing pythons and watching for acid beetles, which tear your skin if you smack them.

This is the kind of stuff you read in books and *National Geographic*! People ask me why I go to those places, and I tell them it's because I want to know where these young Airmen are living, and I want to acknowledge the impact they're having on the mission at large. I want each of them to know how important they are to the central mission of the U.S. Air Force.

We had special operators in the southern Philippines, a dangerous location surrounded by a jungle. Every time we'd go down, the Airmen would say, "Wow! Why'd you come all the way down here?" We just wanted to tell them thanks for what they were doing, and make sure they had received the right training and equipment.

When I traveled with General Schwartz to these outposts, he would listen intently to the Airmen and consult with staff, but he would make decisions based on what he saw and heard rather than what staff members were telling him. Yes, leaders have to be in the Pentagon, on the Hill, and in the Tank, but you also need to know what's going on in the field. As Air Force chiefs of staff, General Schwartz and General Welsh were in the field. They cared about individual Airmen enough to go visit small groups of them in remote locations around the world.

*   *   *   *   *

A good leader is even-tempered and patient. If a subordinate makes a mistake, don't chew them out. Take them aside privately and point out what they did wrong and why.

One time I was in a deployed location where I was just starting another tour through the AOR.[4] I stopped at Al Udeid Air Force Base in Qatar and held an all-hands call. An Airman asked me a simple question about education. I shared my general thoughts on education and suggested they should be going to school, even in a deployed location. If they're away from family and not mowing the grass or doing other chores, they probably have the time. A journalist there wrote a news article and quoted me, and the gist was, "Chief says education essential for growth and development."

A little while later I got a call from General Schwartz.

"Hey, Chief, I just read this article," he said. "Just wanted to let you know I'm not comfortable with education being discussed in theater during combat."

He didn't call to chew me out, and I appreciated that.

"Understood, Sir," I responded. Then I explained my reasoning behind what I'd said.

He heard me out, and the way he approached the situation wasn't a matter of "shame on you"; he simply explained why it wasn't appropriate given the context.

When the shoe's been on the other foot and I'm the leader doing the correcting, I give the person a second chance, or at least an opportunity to make it right. It's important to teach people so they understand their mistake and learn to do it differently the next time.

*   *   *   *   *

Regardless of how tired you are or what group you're speaking to, shake the hand of that one person who wants to greet you, or smile and take a photo with someone who wants a quick shot. Leaders are often on a pedestal, and the higher up you are, the more people will want to meet you. Someone gave me that advice, and told me to remember that this might be a person's only chance to ever meet a Chief Master Sergeant of the Air Force. *My gosh,* I thought. *That's a lot of responsibility.* In fact, I was in the Air Force almost 16 years before I met a Chief Master Sergeant of the Air Force.

If someone has a question for you or wants to make a comment, be available for them. Good leaders always make time for their subordinates. ∎

---

4 Area of responsibility: a pre-defined geographical region assigned to combatant commanders, in which they have the authority to plan and conduct operations.

Lesson Five: Joint and Coalition Forces

There are no problems we cannot solve together,
and very few that we can solve by ourselves.
~ Lyndon B. Johnson

In my civilian job, I send a weekly report to the chief operating officer of our company and others in management, and my salutation is always "Dear teammates." That's how I see those I work with, and that attitude has helped me better understand people and what motivates them to accomplish goals together.

In tech school, I learned that you can be a good operator or a good technician, but unless the whole team is working together, you won't succeed. It's not about one individual—it's about the Air Force as a whole.

The first thing I did when I became Chief Master Sergeant of the Air Force was to meet with the leadership team, including Air Force Chief of Staff General Schwartz and Air Force Secretary Michael Donley. We defined where our focus would be so I could better understand their intent. Then I sat down with my team and established three priorities. One of the things you realize as an engineer is that you can't do it all yourself—a lot of work goes into a project, and you're not going to complete it solo. You have to rely on the whole team.

Who I considered my teammates, however, were not just Airmen. Throughout my military career, I gained a unique understanding of and passion for the joint force as I worked and trained alongside Soldiers, Marines, Sailors, and Coast Guardsmen. When you're with people from all five services who care for each other as teammates and support each other's missions, your work is especially rewarding.

I was fortunate to have early exposure to a joint community. At Fort Leonard Wood, I stood out on the field with Soldiers and Airmen, teaching them how to operate the equipment. I learned the differences between Army culture and Air Force culture, and I understood what Soldiers bring to the fight. We also had Reserve and Guard Airmen in our class, and all components had something valuable and distinct to contribute. I learned how we can work together and what capabilities each service brings. I also became familiar with the different levels of experience, education, and training, and understood how we could share those unique aspects of our culture with each other.

When I was TDY[5] to Kadena Air Base in Okinawa, Japan, I worked with a lot of Marines. My team was remodeling the NCO Academy at Kadena, and my particular job was to redo all the driveways, curbs, and sidewalks. Unfortunately, we didn't have enough equipment to do the job, but I quickly came up with a solution.

I went over to one of the Marine Corps bases and asked, "Hey, can I use some of your equipment to do this job?" The Marines didn't have that type of work when they were in garrison. They trained, trained, and trained, but they didn't have that real-world work

---

[5] Temporary duty assignment

experience—like re-paving driveways. They were very agreeable.

"Yeah, you can have our equipment," they said, "but you have to train some of our Marines." That was a deal I was happy to make.

Just by virtue of being a civil engineer, I interacted with our joint warfighters at a very tactical level. That was an opportunity I've never taken for granted.

A lot of areas lent themselves easily to joint operations, but what we were trying to do was get the force at large to understand and appreciate the whole idea of what a joint partner means. How can we work together, what capabilities does each service bring, and what capabilities can each provide to that warfighter? Those were issues we regularly discussed.

While I was in the seat, we sent our Airmen to the U.S. Army Sergeant Major Academy and the Navy Senior Enlisted Academy, and had them work with the Coast Guard. We brought Marines for the first time into the Senior NCO Academy, and we sent our Airmen to their courses as well. It solidifies the whole idea of a strong Department of Defense, because it's not about one service, or what you can bring individually. It's about what you bring to the fight collectively as a group, and what that team is able to produce with all those assets.

I know our current leadership wouldn't even think of going into combat without our Total Force partners.

*   *   *   *   *

One week, I was at Camp Leatherneck in Afghanistan. My team had gone to see Airmen stationed in a remote site in the CENTCOM[6] theater of operations, and we had just visited the hospital there. Our medics were doing an amazing job. We talked with a lot of the wounded, and with the doctors, nurses, and technicians.

The helicopter landed and I got ready to leave, but someone stopped me.

"Chief, can you hold off for a little bit? Sergeant Major is on his way back from the front line and he wants to see you."

He was referring to the NATO Regional Command, South West Command senior enlisted leader responsible for the entire combat operations in the south, so of course I was happy to wait for him. When he arrived, he said, "Chief, listen. I want you to know, and I want you to relay to every Airman, that my Marines will fight harder every single day because they know that your Airmen are there for them if they get hurt. They know they have the support of combat medics like we have here in Camp Leatherneck."

That meant so much to me, and I shared that with every Airman I talked to. It wasn't just the combat medics who were valuable, it was every single Airman we had out in those locations.

That Marine sergeant major was Mike Barrett, who became the Sergeant Major of the Marine Corps while I was the Chief Master Sergeant of the Air Force. We were a strong

---

6 The U.S. Central Command, or CENTCOM, is a theater-level Unified Combatant Command of the U.S. Department of Defense. Although the CENTCOM AOR includes countries in the Middle East, parts of northern Africa, and Central Asia, its main headquarters are located at MacDill AFB in Tampa, Florida.

team, and joint teamwork goes to the heart of who we are. We are Americans first, and we'll do whatever America needs of its Armed Forces.

Since 9/11, military leadership won't even think of going into combat without our Total Force partners—not just Active, Reserve, and Guard, but also civilians. Much of my training at MacDill AFB came from the civilian workforce. A civilian by the name of Jack Hood taught me everything I know about operating a TD25 dozer, which made me a much more valuable member on a rapid runway repair team as we prepared and continued to stay ready in Korea.

When I was the senior enlisted leader and advisor to Admiral Keating, he didn't separate those of us on his staff according to what service we were in. It didn't matter—we were all on Team Keating. That's what he called us. He believed it took a whole team to make a mission happen.

I sometimes called Admiral Keating "Airman" because of his approach to the joint forces. He was unique—a naval officer at the very top who considered all of us in the service as equal players on the field. I picked up on that attitude, so when I went into the position, Team Roy was what I called my staff.

*　　*　　*　　*　　*

In addition to the teamwork of joint forces within the U.S. Armed Forces, our partnership with coalition forces was also crucial in our efforts around the world. In the Pentagon, we regularly discussed what our coalition partners brought to the table, and what we could offer them.

During my tenure at the Pacific Command, Admiral Keating charged me to be the advocate and champion for the enlisted development of our coalition partners in theater. I was able to travel to many locations throughout the Pacific Theater and work with our coalition partners. I was fortunate to have the opportunity to get to know a lot of them at different levels and be able to share what we have as the U.S. Air Force and Department of Defense on growth and development.

When I worked with some of our allies, our joint partners and our coalition partners, it was easy for me to explain this idea of professionalism and mutual respect. In some cases, however, our model of teamwork was almost incomprehensible; many countries have a caste system and a conscript force, where everybody has to serve in the military at a certain age for a certain period of time. The U.S. all-volunteer force is, of course, quite a different environment. Our model of education, training, and experience is an enormous asset to our commands and a positive example for our coalition partners.

During my time in the seat, we were able to put a couple of our senior NCOs into the New Zealand senior leader course, and they got full credit for the U.S. Air Force Senior NCO Academy. That says a lot about the level of professionalism of our force. The Air Force strongly supports the idea of joint and coalition partnership, and the Airmen are enthusiastic about it. It's beneficial to the organization, the individuals, and the commander when they call upon us.

I look at what we did with the Inter-American Air Forces Academy (IAAFA) in San Antonio, Texas. One time when I was visiting, I asked the NCOs, "Why is it we send our

NCOs and Airmen to Colombia or other places in Latin America, we teach the course, and we bring them to us to teach, but we don't attend the course with them?"

So they did just that. Our Airmen were able to go down range and go to school with their Colombian partners. They are going to be warfighters together and learn side by side. We've also sent some of our Spanish-speaking Airmen from across the Air Force back to IAAFA to sit with many other nations and attend the exact same course, particularly the NCO Academy. That highlights the level of professionalism of our Airmen, and it certainly shows that the institution sees value in what it does for our Airmen and the combatant commander.

Today, the partnership aspect with other countries is light years ahead of where we were 20 years ago. As a young senior Airman, I was stationed at Osan, South Korea, and traveled around the entire peninsula. I was TDY to a lot of those locations for extended periods of time doing different projects. We knew those were Korean air bases, but we never even knew the Koreans. We never associated. I look back at that now and I think, *Wow, how much more potent of a force could we have been if we had worked with them and learned from them?*

I went back four years later to Kunsan Air Base and got to know many of the airmen serving in the Republic of Korea Air Force. We had a good partnership; we helped them with things they needed, and they shared their understanding of the peninsula with us. Fast-forward to 2007, when I was the Pacific Command senior enlisted leader, and it was completely different across the entire Pacific theater of operation.

When a partner country is going through a complete defense reform, and they ask the United States for assistance, that is a great compliment to our military. Our intention was to raise their level of professionalism so the partner country could be the first in the game. Why shouldn't the home team be the first defense?

Members of the U.S. Armed Forces grow from being around coalition partners and joint partners. You can see a complete difference. Those are the things leaders have to continue if we are going to be of value to our combatant commanders who fight the wars.

* * * * *

I sat on a lot of boards and in a lot of meetings during my time as Chief Master Sergeant of the Air Force. Many times I wasn't the decision maker, but I was there to provide advice or input. Of course, as with most meetings there was a wide range of personalities, and some people just wanted to be heard. Sometimes the input was just, "We've heard that already; we need to move on."

My approach has always been to speak up only when I have something to say. Whether I agree with it or not, I'm going to tell you. But I don't need to be heard by the rest of the group. I've been told that I'm quiet in meetings, but whenever I did speak up, the Secretary, the Chief of Staff, and Admiral Keating all listened to me. They knew if I spoke up, I had something to say.

It's good board "etiquette," if you will, to stop talking in a meeting if you're not adding any value to the subject. Just move on. ■

Lesson Six: Lead by Example

If your actions inspire others to dream more, learn more,
do more and become more, you are a leader.
~ John Quincy Adams

To do my job well as Chief Master Sergeant of the Air Force, I wanted to learn about the special ops forces. I've always said that if you have a better understanding of the people and their missions, you'll make sounder decisions. The Air Force Special Operations Command invited me to spend an entire week with them, and it was fascinating. One of the units we visited was the 24th Special Tactics Squadron, an Air Force component of JSOC (Joint Special Operations Command). Of course, I can't divulge exactly what I did, but they took me through the whole gamut. The culminating experience—I'm allowed to disclose this—was jumping out of a perfectly good airplane in a tandem parachute jump.

When I broke the news to Miss Paula that I was going to jump, she was not happy.

"Why are you doing this?" she demanded. "What are you trying to prove?"

"I don't have a choice!" I told her. "I *have* to do this."

"Yes, you do have a choice!" she retorted. "You're the Chief Master Sergeant of the Air Force! You don't have to do this!"

"Dear, you don't understand," I said. "I really have to do this. If my Airmen are jumping out of planes, I've gotta do it too."

"Well, OK," she said, giving in. "But then you call me as soon as you hit the ground."

"When I hit the ground?" I repeated, raising my eyebrows. "Interesting choice of words!"

They paired me up with one of my heroes, a man they called the Flying Hawaiian. Once we got up to the right altitude and I got to the edge of that ramp, I started getting a little nervous. I was glad there was someone behind me who was jumping, because I probably would have backed away otherwise. I jumped out of the plane, and everything functioned normally.

If you look for it, every experience can teach you a lesson. For me, the lesson from my parachute jump was this: if you're going to be in a leadership role, you can't just say what you're going to do, you've got to *show* people you'll do what it takes to understand what you're asking them to do. If you're not willing to do it, why in the world would they want to?

As always, there's a reverse example. A young lieutenant I worked with would reprimand the Airmen because they didn't have their shoes shined or their uniforms were wrinkled. I pulled him off to the side, behind closed doors.

"Look at this!" I said, gesturing at his entire uniform. "How can you bark at them when you showed up to work late, and look at yourself! Your shoes aren't shined, and your uniform looks like you slept in it. You will not get up in front of them again and do that."

That lieutenant did a complete 180 after our discussion. He took my advice to heart and made sure his uniform was sharp and his example impeccable. Because the Airmen saw that he practiced what he preached, they respected him. The mutual respect we had for one another ensured an exceptional flight.

* * * * *

After my jump with the 24th STS, I was scheduled to speak to the entire flight. I went into the restroom to collect my thoughts, and took a couple of deep breaths. I had been incredibly impressed with each one of the Airmen, and now I had the humbling opportunity to speak to them. *What in the world am I going to tell these warriors?* I thought. *How could I possibly motivate them? They're the ones who motivate me.* These are the men and women who are putting their lives on the line, who are stepping out first when the commander in chief makes a critical decision and sends our troops into harm's way.

I don't remember what I said, but I do remember feeling very humbled. I was overwhelmed with a sense of awe in front of those courageous Airmen.

I don't care how high up in an organization you are, you have to be humble about the responsibility of leading—whether it's a group of warriors or a group of civilian employees. Leaders need to be an example of humility and respect for those they lead.

I used to think leadership was inherent, but you learn certain traits on the job. I'm still in the learning process.

* * * * *

When I was in high school, there was no discussion of college. It wasn't even an option, because my mom couldn't afford it. Same for Miss Paula's parents. It was only later in life that I got my degrees, thanks to the Air Force and some great supervisors along the way who helped push me.

When I had been at my first duty station, MacDill Air Force Base, for about a year, I got an alarming letter from the education office.

"Airman Roy," it read, "your reading comprehension is not high enough."

Uh oh, I thought. *What do I do now?*

The letter explained that I needed to take some courses to improve my reading comprehension. Not only did my scores jump dramatically after the courses, but I eventually received three associates degrees, a bachelor's degree in engineering management, and a master's degree in human resources management. The Air Force took an 18-year-old kid with a reading comprehension problem from a little town in Michigan and made it possible for him to receive five degrees, write a course for the engineering program, and become Chief Master Sergeant of the Air Force.

With the right education, you can overcome a lot of things.

Most of the colleges I attended were at Fort Leonard Wood in the Missouri Ozarks. My education not only gave me an abundance of technical knowledge, but also helped me improve my verbal and written communication. It gave me the motivation and desire to continue to improve myself and gave me the tools to prepare for a presentation. I know what and how to study to understand a subject, how to determine the audience's needs, what message I need to convey, and what potential questions people may have. A college education must be a priority for any leader; you have to show the path for others to follow.

*   *   *   *   *

Though I started at the bottom, a walk-up Airman Basic excited to operate large machines, I climbed the ranks to become a Chief Master Sergeant of the Air Force. The winding path that got me there brought me in contact with leaders of all abilities who have taught me what it means to be an effective leader as well as an effective member of the team. My learning by experience and observation has granted me the opportunity to become a powerful leader myself. Just like the leaders who came before me, I know that by implementing everything I've picked up along the way, I've had the privilege to influence more men and women than I can count.

I believe we have the absolute best-trained, best-equipped, best-led, all-volunteer, professional, combat-hardened Airmen we have ever had. I sleep safe and sound, knowing that our Airmen are on sentry. If we look at the legacy of all those who have gone before us, we have built it up to the point where the next generation is going to be even better. I couldn't be more proud of the Airmen who serve today and the Airmen who will serve in the future. ■

Lesson Seven: Transitioning

Change is the law of life. Those who look only to the past or present
are certain to miss the future.
~ John F. Kennedy

In the military, we call transitioning to the civilian world "the next chapter." Any change seems daunting after nearly 31 years of being privileged to wear the cloth of our nation, but transition wasn't as difficult as I had imagined—probably because I chose to stay in the defense industry, which keeps me engaged with service members. However, the learning curve was steeper than I had anticipated. I can group what I learned into three segments:

1. Learning from the past

Experience in the military taught me to be prepared. That skill helped me as a civilian, because I needed to be ready for the unexpected. Miss Paula and I chose a location to retire to and thought we'd live there forever; we didn't plan on moving just a few years later. But when my company asked me to serve in a position with increased leadership responsibilities, I was up for the challenge—even though that meant moving to another state. I had learned in the Air Force to be adaptable and resilient.

My mentors shared their transition successes and challenges and gave me ideas on how to avoid the same pitfalls. Find a mentor or coworker who will share their experiences with you. I reached out to many leaders, some with similar backgrounds and some with backgrounds very different from my own. Much of the advice I've received was to complete my education and certifications.

I consulted with my family during the difficult decision to retire. A supportive family is a treasure to be cherished and honored. I always tell Airmen they need to thank their families often for their many sacrifices.

Make sure you are financially fit to make the transition. Unanticipated expenses are a given: have an emergency fund and save enough to get you over the transition period. Let your experiences guide your saving and spending behaviors.

2. Living in the present

The same teamwork that made your military units complete their missions must also exist in your corporations and organizations. Your teammates depended on you then; different teammates depend on you now. Remember that you can only succeed as a team, and you must continue to be a team player.

Look for opportunities to take the initiative and use the leadership skills you've acquired. Most of your civilian peers have never attended leadership courses or been in situations that required those skills.

Realize it's OK to not know everything; learning is a lifelong event. Are you willing to take on that new position? What skills do you need to continue to hone? Be willing to push yourself beyond your comfort zone.

Core values are our moral compass and guide us to do the right thing. Only associate with companies or organization that espouse the same ethical standards you had in the military.

3. Live looking forward

Anticipate and pursue new opportunities for you and your family to grow. As I watch my children become adults and move on to colleges and into the workforce, I'm reminded that this is the cycle of life. Their future excites me and forces me to think of ways for them to serve and for us to continue to flourish as a family.

I consider myself blessed for what we have and what we've been able to accomplish. I'm deliberate in my assistance to help those I believe in. By mentoring others around you to succeed, you will also continue to grow.

*　　*　　*　　*　　*

After retiring from the Air Force in April 2013, we moved to Summerville, South Carolina. I began working for Scientific Research Corporation (SRC), an advanced engineering company providing innovative, cost-effective technology solutions for air, land, sea, and cyberspace applications.

I'm also a military advisor for Veterans United and serve as a board member for the Armed Forces Benefit Association, Air Force Association, and the Air Force Sergeants Association. These organizations are focused on serving those who serve, being the voice for aerospace power and the Air Force family, and educating military members and families about their veteran benefits. I'm very grateful to have the opportunity to be a part of these four outstanding organizations; they keep me connected with the men and women of our military.

I'm passionate about professional organizations that foster the growth, development, and support of our military members and their families. I am connected to many of these, including the American Legion, Veterans of Foreign Wars, and Armed Forces Communications and Electronics Association International.

In my spare time, I enjoy restoring old vehicles with my son Caleb. I've especially enjoyed spending time together while teaching him a worthy skill. My son Colby and I enjoy intellectual discussions on leadership and politics. He is remarkably mature, and both of us think more deeply and research specific topics when we're together. Miss Paula is still the love of my life, and we continue to have a special relationship of both love and friendship. We enjoy talking about the day's activities and our sons' futures, and we still discuss ideas on how we can continue to support our military members and their families. ■

TOP: Tech Sgt. Lance Davis and Staff Sgt. Roy, Fort Leonard Wood, Missouri, Training Area 244, Dozier Phase, Oct. 1989.
MIDDLE TOP: CMSAF Roy addressing Airmen deployed to Middle East, June 2010.
MIDDLE BOTTOM: Miss Paula and CMSAF Roy, "A Walk to Remember," Pentagon, Jan. 2013.
RIGHT: CSAF Welsh and CMSAF Roy addressing Airmen, Malmstrom AFB, Montana, Nov. 2012.

TOP: CMSAF Appointment Ceremony; CSAF Schwartz, Admiral Keating, and CMSAF Roy, Bolling AFB, Washington, D.C., June 2009.
ABOVE: Miss Paula and CMSAF Roy, Air Force Memorial, Arlington, Virginia, Memorial Day 2011.
LEFT: Roy Family, Hickam, Hawai'i, June 2008.

ABOVE: Wedding day, Monroe, Michigan, 1982.
RIGHT: Celebrating 12 Outstanding Airmen of the Year with SECAF Donley and CSAF Schwartz, National Harbor, Maryland, Sept. 2009.

TOP LEFT: With my dear mom, Dora Roy, Michigan, 1967.
TOP RIGHT: Our mentors, April 2016.
TOP RIGHT BOTTOM: Office of CMSAF, Sept. 2012.
ABOVE: Friends and family help us celebrate retirement, Andrews AFB, Jan. 2013.
LEFT: Visiting Nangahar Provincial Reconstruction Team in Jalalabad, Afghanistan, Nov. 2009.

Denise M. Jelinski-Hall

D enise M. Jelinski-Hall served as the Senior Enlisted Advisor to the Chief of the National Guard Bureau from February 19, 2010, to June 6, 2013. She holds the distinction of achieving the highest position ever held by an enlisted woman in the history of the U.S. military.

Chief Jelinski-Hall was born in Little Falls, Minnesota. She reported to Lackland Air Force Base, Texas, for Air Force Basic Military Training on November 14, 1984, and completed Air Traffic Control Training at Keesler AFB in Biloxi, Mississippi. She was assigned to Offutt AFB, Nebraska, as an air traffic controller in July 1985. In April 1987, she transferred to the California Air National Guard and served as a ground radio operator. She and her family moved to Hawaiʻi in 1990, where she joined the 297th Air Traffic Control Flight at Barbers Point, Hawaiʻi, as an air traffic controller, mobile tower chief controller, package chief, and superintendent for the ATC section.

In 2002, Chief Jelinski-Hall transferred to the HQ 201st Combat Communications Group, Hickam AFB, Hawaiʻi, where she served as a combat airspace manager. In 2004, she was selected as the first female command chief for the 154th Wing, Hawaiʻi Air National Guard. She was designated command Chief Master Sergeant for the Hawaiʻi Air National Guard in 2006. Shortly after, the adjutant general of Hawaiʻi selected her to serve as the state senior enlisted leader for the Hawaiʻi National Guard (Army and Air), the first woman in that position. She held both positions simultaneously until Gen. Craig McKinley appointed her as his senior enlisted advisor (SEA) for the National Guard Bureau, the first member of the Air National Guard and first woman to serve in that position.

Chief Jelinski-Hall's responsibilities included advising the chief on all enlisted matters affecting training, utilization, health of the force, families, and enlisted professional development of more than 415,000 National Guard Soldiers and Airmen in the 50 states, three territories, and the District of Columbia ("the 54").

While assigned to the Pentagon, Chief Jelinski-Hall served on numerous executive-level committees representing the National Guard Enlisted Corps. She travelled across the 54 and overseas, visiting with National Guard Soldiers, Airmen, and their families. She brought back critical information and feedback from the field to improve policies, programs, and quality of life for all Guardsmen. She represented the National Guard at the White House and at senior-level meetings and events with the Secretary of Defense and Joint Chiefs of Staff. Chief Jelinski-Hall spoke to diverse audiences throughout the world and routinely engaged with members of Congress, their staffs, and senior industry leaders on Capitol Hill.

Her major awards include the Defense Superior Service Medal, Legion of Merit,

Meritorious Service Medal, Hawai'i National Guard Distinguished Service Order, the Minuteman Award, and the Department of Defense Trailblazer Award, National Guard Bureau. In 2014, she was a Top 100 finalist for the John C. Maxwell Transformational Leadership Award; the next year, she was a Top 30 finalist. In 2015, she earned a bachelor's degree in business administration from Trident University International.

Chief Jelinski-Hall served as director for Military Child Education Coalition (MCEC) for a three-year term. For two years, she was the director of Colorado Air National Guard membership, and she currently volunteers with the Colorado Springs chapter of Employer Support of the Guard and Reserve (ESGR). She also serves as a trustee on the board of directors for United Through Reading (UTR) and as secretary on UTR's executive committee. She is a military advisor at Veterans United Home Loans (VUHL).

Chief Jelinski-Hall is the author of *From the Prairie to the Pentagon*, an autobiography chronicling her upbringing and military career. She speaks at military and civilian events on leadership and the importance of professional and personal growth and mentorship.

In December 2017, she delivered the commencement address for the winter graduation ceremony at Hawai'i Pacific University, where she received an honorary Doctor of Humane Letters, a degree presented to those who have dedicated their lives to the service of others.

Denise and her husband, Gary Hall, have four children and six grandchildren. They live in Colorado. ■

Introduction: Life on the Prairie

You will never plough a field
if you only turn it over in your mind.
~ Irish proverb

I was born and raised in a rural farming community in Central Minnesota, the third of six children. My family lived on an 80-acre farm with beef cattle, occasional pigs and chickens, and seemingly endless chores: bailing hay, picking rocks, fixing fences, canning, mowing grass, and caring for a huge weed-free garden were just a few responsibilities we all shared. My siblings and I knew how to work hard from a young age, but we also played and had a lot of fun together.

My mom was a kind, sweet woman, a traditional farm housewife who ran the household and our garden with a gentle hand. We always had food to eat, a warm bed, and clean clothes, but there was no extra money for frivolous things. Devout Catholics, my parents raised us with traditional family values and grounded us in our faith: honor the Lord, respect your elders, be honest, treat everyone with dignity and respect, and don't do anything to tarnish the family name.

My dad had served in the Korean War as a Soldier in the U.S. Army. Although he never spoke about what it meant to serve his country or what he did in the war, I was always aware that he had served, and I was proud of him for defending our country. After coming home from Korea, he joined the Minnesota Army National Guard and rose through the ranks, becoming a first sergeant and ultimately retiring as a chief warrant officer 2. My father fit the profile of a crusty old first sergeant: tough as nails (some said he "spit tacks for breakfast"), but fair. He was well known and respected in our community. Although my parents didn't often show affection, my siblings and I knew they loved us.

We grew up in an era where your word was your bond and a handshake sealed the deal. We learned what it meant to work hard, never quit, commit to a job, and do it right the first time. I didn't know it then, but these values would affect my life and success from that point on. Details mattered. Our parents showed us how to do something once, and we were expected to do it right every time thereafter. There was no complaining or talking back. We executed Dad's orders like good little soldiers.

We worked six days a week, but Sunday was the Lord's Day. Even if the skies promised rain on Monday, we knew it was better to let hay rot in the field than to dishonor the Sabbath. We lived about a quarter-mile down the road from my dad's parents, Grandpa and Grandma Jelinski, who reinforced our parents' values. They were a big presence in our lives and an integral influence. Somehow we always knew when Grandma was baking homemade bread and sweet treats. We'd either walk over or ride our bikes, and more often than not we'd return with a warm loaf of bread cradled gently or balanced on our handlebars.

Until the sixth grade, I attended a one-room country school about a mile and a half from our farm. Our school had one patient teacher, Mrs. Boros, who taught all subjects and all grades, first through eighth. There were separate outhouses for the boys and girls, an oil-burning stove, and no telephone—a lot like the Ingalls' schoolhouse in *Little House on the Prairie*. After I finished fifth grade, the country schools closed and we were bused to the city schools.

Mrs. Boros had done an admirable job of juggling so many subjects and grades, but there was simply no way she could give all of us more than a few minutes of daily instruction in each core subject. As a result, I felt poorly prepared for sixth grade. I was placed in most of the lower-level classes—nicknamed "the dummy classes"—and I had to work long hours to master the subjects. Fortunately, I knew how to work hard from growing up on the farm.

Getting good grades wasn't enough to fit in with the "cool" city kids, though, so I decided to join the school band. I had been playing the guitar, but the band didn't take guitarists, so I asked my parents if I could play the drums. Dad said no—he played the drums, and one drummer in the family was enough. My second choice was a clarinet. My parents agreed, as long as I committed to practicing every day for 30 minutes. No exceptions! Until I left home at 18, I practiced half an hour on the clarinet and half an hour on the guitar, every single day. That was an early and lasting lesson on the importance of commitment.

The morning after high school graduation, I started working as a seamstress at a clothing manufacturing plant in town. I disliked the job immensely. I didn't have many other options, though—I couldn't afford college, and I had no marketable skills. I didn't even know how to sew (nor did I want to learn), but my oldest sister was already working at the plant, so it seemed like the easiest place to get a job.

In my heart, I knew I was capable of doing something more. My friends were all going off to college, and deep down I resented having to work at a clothing plant. But my blue-collar upbringing had trained me to be happy and grateful for any kind of job, so I simply accepted that this was my lot in life. I saved money by living at home, and bought a 1967 Ford for $200 to get me to and from work.

A year later, my high school boyfriend and I got married, and I got a new job as a bookkeeper in our town's local bank. My boss noticed how hard I worked, and soon I moved to the teller line, then became the head teller in charge of the vault. I took pride in balancing my drawer every night. My attention to detail paid off, and soon I was given additional responsibilities and trained as the backup loan teller. I was earning about $600 a month. I enjoyed the banking environment, but soon discovered that without a college degree I would not advance very far. Two years later, my marriage unraveled and I found myself divorced and on my own. I was 22.

One day, about five years after I began working at the bank, one of my regular customers came up to my window. Staff Sgt. Joyce Madsen, a member of the Minnesota Army National Guard, said something that changed my life.

"I see so much more potential in you than this, Denise," she said. "Why don't you join the Air Force and get out of this town?"

I was intrigued, and when she offered to take me to the Air Force recruiting office in neighboring St. Cloud, I impulsively said yes. The recruiter did a phenomenal job selling the United States Air Force. I didn't know a thing about air traffic control, but when he talked about the field as a career, it sounded cool. When he described the honor of serving my country, along with the benefits, I was hooked. I could see myself doing something prestigious and exciting that didn't require a college degree. On top of that, the recruiter said I'd be able to pursue higher education in the Air Force.

I decided to join on the spot. Six weeks later, I had sold my car and furniture, resigned my position in the bank, stored my personal things at my parents' house, and was off to the wild blue yonder. I leapt at the chance to make something of myself, and never looked back. ■

Lesson One: Trust Your Intuition

It is by logic that we prove,
but by intuition that we discover.
~ Henri Poincaré

Joining the U.S. Air Force felt right. It was the first time I had truly listened to my intuition, and everything inside me was saying this was the right direction for my life. Shortly after I made my decision, my recruiter called and explained that the air traffic control field was full. I could either choose a different career field or go in "open general," which meant the Air Force could put me wherever they needed me the most. My intuition told me it would all work out, so I went with open general. I was all in and ready to commit.

Trusting my intuition from the very beginning of my career became something I relied upon heavily. In 1986, I met a handsome Marine named Gary Hall. We had only known each other for three weeks and had gone on five dates when he asked me to marry him—even more surprisingly, I said yes without hesitating. Logic might have told me that was crazy, but intuition told me it was right.

We were married six months later. That was more than 30 years ago, and I've never regretted listening to that inner sense that I was on the right path.

Whenever there was a big decision to make, I took time in prayer and listened to what my instincts were telling me. Sometimes those feelings were at odds with what seemed reasonable or rational, and my heart and brain had to fight it out. But I've never gone wrong following my intuition. In 2001, I was serving in the 297th Air Traffic Control Squadron at Kalaeloa Airport (formerly Naval Air Station Barbers Point) on Oahu, Hawai'i, when I had the feeling I needed to move on. After 12 years in the squadron, and as a senior master sergeant, I believed a change was best for the organization and for the air traffic controllers I led. It was their time to shine and take on new responsibilities so they could grow stronger as leaders. Besides, I was hungry for a new challenge and wanted to develop further as a leader and as an Airman.

One evening, I told our former chief air traffic control officer that I would be leaving the unit and my full-time technician job after the next inspection, and I would be looking for a drill-status Guardsman position. He was astonished that I would give up a senior full-time position without having something else lined up. Not only that, but as a senior master sergeant in the ATC squadron, I was an E-8 assigned to an E-9 position—so if I stayed in the squadron and continued being a good Airman, I most likely would be promoted to chief master sergeant.

I knew that leaving before I was promoted didn't make sense from a logical point of view, but I knew it was the right decision for me. I trusted that intuition even after I was offered a lateral transfer to the Headquarters 201st Combat Communications Group (CCG), where there was no possibility of promotion to chief. I knew that senior master sergeant was the highest enlisted position at the headquarters, but if I was capped as an E-8, I was okay with that. Of course, most enlisted Airmen aspire to become a chief master sergeant, but sometimes the timing, opportunity, or path to chief do not exist. For me, it was about service and the greater good of the organization. I had prayed a lot before my decision; I trusted in the Lord, and intuitively knew it was the right thing to do.

I accepted the offer and transferred to the HQ 201st CCG into the combat airspace manager position. A year later, when I was selected as command chief master sergeant for the 154th Wing on Hickam Air Force Base (now Joint Base Pearl Harbor-Hickam), my earlier decision started to make sense. If I had stayed in the Air Traffic Control Squadron, I might have missed this opportunity. Perhaps there was something I needed to learn at Headquarters before I could take the next step to becoming a command chief. In any event, I was glad I had trusted my intuition—being assigned to Hickam AFB opened many new doors for me.

Not too long after that, Steve Jobs gave the commencement speech at Stanford. He told the graduates to trust that they knew where they were going:

> You can't connect the dots looking forward; you can only connect them looking backwards. So you have to trust that the dots will somehow connect in your future. You have to trust in something—your gut, destiny, life, karma, whatever. Because believing that the dots will connect down the road will give you the confidence to follow your heart even when it leads you off the well-worn path, and that will make all the difference.[1]

Looking back, I've been able to connect the dots in my life. Over the course of nearly three decades, there were countless times, after many prayers, that I trusted in God's plan for me when it made no logical sense, and it served me well. By doing so, I was exposed to tremendous opportunities and experiences, and I met incredible people along my journey. As a small kid on the farm, I never dreamed these things were possible.

When I was trying to decide whether or not to apply for the position of senior enlisted advisor, National Guard Bureau, I prayed about it and listened to my instincts. My family and I had lived in Hawai'i for nearly 20 years, so this would be a big jump outside our comfort zone. Initially, I was uncertain about applying. Once again, though, I listened to what my inner voice was telling me, and I applied. The calling to serve at a higher level and make myself available to senior leadership was never clearer.

Cultivate and trust that inner voice or the feeling deep inside you. Many times, it's opportunity knocking and calling your name. ■

---

1 Steve Jobs, commencement speech at Stanford University (June 12, 2005).

Lesson Two: Embrace Challenges and Take Risks

A true leader has the confidence to stand alone, the courage to make tough decisions and the compassion to listen to the needs of others. He does not set out to be a leader, but becomes one by the equality of his actions and the integrity of his intent.
~ Gen. Douglas MacArthur

Challenges are inescapable and part of everyone's life, but that doesn't make them any easier. My first test as a new Airman came two days after arriving at Basic Military Training (BMT), when I heard my training instructor announce, "Jelinski, you're going to be my dorm chief." I immediately responded the way I'd been trained.

"Yes, Ma'am," I said dutifully, although I had no idea what a dorm chief was or what was required of one. It sounded like an assignment nobody wanted, but it wasn't like I had a choice.

Being the dorm chief for my flight presented many challenges and stretched me to my limits. I was responsible for overseeing the flight whenever our TI was not present. Our flight was somewhat dysfunctional, so there was much to do. I spent countless hours helping my sister trainees, writing painstakingly detailed reports, ensuring daily assigned tasks were completed, and standing watch over those at risk for suicide. Some women struggled with deep emotional and physical issues. I quickly realized I had to find a different voice deep within me—literally. I was enormously uncomfortable with raising my voice—even my stern dad had never yelled at me—but I surprised myself by how forcefully I could bark orders when I had to. I learned to be direct when issuing tasks and to be firm and strong to ensure compliance.

That first leadership opportunity in BMT required me to think, find solutions, develop teamwork, build trust within my squad leaders, and quickly understand and rely on the chain of command, both up and down—I had to rely on my subordinates as much as I did on my superiors. I had a steep learning curve, but mastering it gave me the confidence to take on the next challenge: Air Traffic Control (ATC) school.

Air traffic control did not run in my blood. Unlike some of my classmates, I knew nothing about it, and studying it was like learning a foreign language. On top of that, I had been out of high school for almost six years, so my study habits were rusty. I had to work three times harder than many classmates to absorb the mammoth amount of material. It was difficult for me to discern what was important and what was peripheral, so I spent an inordinate amount of time memorizing large amounts of information. I spent virtually every waking moment studying—even at 0300 under my blanket with a flashlight, or on weekends when others went to the beach or to the movies. I refused to fail just because I hadn't put in every ounce of effort and time into mastering the subjects. Determined, I gave up everything non-essential to learn the ATC material and prepare for block tests.

As the superintendent in the air traffic control squadron, I had to contend with senior leaders who didn't communicate effectively and had personal issues that affected the entire unit. Rather than wait for guidance that often never came, I just had to figure it out on my own. I came to realize that challenges are merely opportunities for growth; these difficult and uncomfortable experiences made me grow, stretch, and think differently and broadly to find solutions.

You have to look at the challenge or obstacle from a different perspective. How do you break through a barrier? My experience has been that most often there is a way around, under, over, or right through the middle of whatever challenges lie ahead of you. I don't mean to imply that one should circumvent the rules, just look at situations and problems from a different angle. The answer is typically just under the surface or around the corner; the following principles helped me find it.

1. Acknowledge the challenge

Booker T. Washington, one of the foremost African-American leaders of the late 19th and early 20th centuries, was no stranger to challenges. He said, "You measure the size of the accomplishment by the obstacles you have to overcome to reach your goals." Recognizing the challenge is half the battle; having the humility and self-awareness to understand your limits and weaknesses will help you move forward. When I needed to complete Intermediate Algebra for my associate degree from Community College of the Air Force, I knew I was in big trouble. I had struggled with math ever since grade school, and here I was, nearly 30 years after high school, still struggling. I realized I would need a lot of extra help, so I hired a young Airman to be my tutor.

For ten weeks, on every weeknight and for countless hours on the weekends, this patient Airman helped me understand algebra. More often than not, we were the last people in the Hickam AFB library when it closed at 9:00 p.m. I dreaded class night; it was humiliating and extremely frustrating. Tests were an absolute nightmare.

At the end of the course, however, with every possible extra-credit point, I got an 89.4. I was elated. That tutor cost me more than the class, but without his assistance, I would not have passed Intermediate Algebra. I will be forever grateful for his time and endless patience!

2. Find a resource

Whatever challenge you're facing—career, financial, relationship, spiritual—you can always find a resource to provide assistance. Don't be too proud to ask for help. Acknowledge that you will need help with some challenges, and be humble enough to give a lot of credit to those who pull you through.

3. Embrace the opportunity

Everyone has challenges, regardless of age, rank, educational background, or position. What you *do* with those challenges, however, defines who you are. Do you buckle when you hit a roadblock? Do you handle problems grudgingly, all the while resenting the people or circumstances that created them? Or do you embrace challenges as an opportunity to learn, grow, and excel?

You don't get emotional victory points for dealing with challenges while complaining about them. Embrace and welcome them! Like attitude, deciding to grow in your professional and personal life is a choice. As your career progresses, you'll find obstacles, deployments, and family challenges along the way. These trials help us gain experience and perspective. If you want to become a strong, effective leader, you must embrace the challenges that lead to growth.

4. Keep going outside your comfort zone
As an introvert, I needed to cultivate a more outgoing and engaging personality to become a more effective leader. To grow, I had to get comfortable with the uncomfortable. Not only did I have to get out of my comfort zone, I had to embrace what I feared most: failing and embarrassing myself in front of others. One of the most painful things I did as a young leader was to volunteer to speak in public. I was nervous about making mistakes or embarrassing the organization or myself, but I did it anyway. I came to understand that a leader must be comfortable with public speaking, and the only way to get there is to do it frequently.

As I continued on my path of self-development and personal growth, I had many opportunities to attend senior leader conferences. I listened carefully to the words other leaders used, and I watched how they carried themselves and how they interacted with people.

A bit of a wallflower, I was more comfortable staying in the shadows, but I clearly saw the importance of engaging with people. Initially, my interactions felt forced and awkward, but the more I engaged with others, the easier it became. As with any opportunity that required me to stretch beyond what was comfortable, I gained confidence the more I put myself out there.

One time I volunteered to co-emcee our annual awards banquet, which was a responsibility well outside my comfort zone. Although I'd been in Hawai'i for a few years, it was still difficult for me to pronounce many of the Hawaiian, Samoan, and Filipino names. I spent hours and hours practicing, and I even wrote the names out phonetically. The program was about to begin when a colonel handed me a new list: people who would be accepting awards on behalf of the Airmen not in attendance. So much for my careful plans! I apologized in advance for my mispronunciation, but there were several snickers in the audience when I stumbled over some of the names. Having planned, prepared, and practiced, all I could do was my best.

5. Be bold—don't take no for an answer
Many times, the challenge I faced was getting a solution approved. This was a huge frustration for me. Often I was simply told, "No—you can't," with no rational explanation. Those roadblocks just made me more determined. Ask questions, read the regulations, seek information and assistance from HR and the employee handbook, ask mentors for counsel, and conduct research to find the correct answer. When someone says *no*, leave no stone unturned to find a way to get to *yes*.

Instead of asking "Can I?", ask "*How* can I?" Assume it can be done. Look things up, take control of your career, leave nothing to chance, and don't let anyone else determine your success. If "no" turns out to be the final answer and a door is closed, look for the open window!

6. Take strategic risks
Effective leaders are not afraid to take a risk. Most senior leaders—whether in the

military or the private sector—did not follow a straight path to the top. It's important to get a broad breadth of experience and look for opportunities to grow, but that doesn't necessarily mean a promotion. You might have to serve in a different capacity and position, which could be through a lateral transfer or a deliberate decision to take a step back to gain another skill set.

Admittedly, that can be risky. Some people might judge you and assume you've abandoned moving up the career ladder or you can't cut it. Giving up a senior full-time position without having something else lined up was a big risk for me—I had to take a leap of faith and jump, trusting that I'd land safely. And if not... I would figure something else out.

Strategic risks don't always result in success, of course, but I believe the cost-benefit analysis shows that the potential reward in professional and personal growth is worth it. An effective leader can't be afraid to make those hard decisions.

7. Don't fear failure

The fear of failure—of making mistakes that might jeopardize your career or just make you look foolish—can be paralyzing. For me, the silver lining was that fearing failure propelled me to succeed and do my best in every endeavor. The fear of failure had haunted me ever since I was put in the dummy classes in sixth grade, and it wasn't until the last decade of my career that I understood the power that particular fear continued to have over me.

About five years before I retired, I attended a three-day leadership course. I was the Hawai'i National Guard senior enlisted leader at the time. One of the last exercises we had to do was break a board on which we wrote our biggest fear. Although my anxiety about failing didn't consume me anymore, it still lurked just under the surface, so I wasn't surprised when it was the first thing that entered my mind. Without any hesitation, I wrote in big bold lettering: FEAR OF FAILURE.

The next part of the exercise was to break the board with an arm chop. By breaking our board, we would symbolically let go of the fear that had the potential to cripple us or, at the very least, hold us back. I held tight to my board and did not want to take a turn. Wasn't the fear of failure part of my identity? Wasn't it what had helped me achieve the successes I'd experienced thus far?

Nevertheless, when everyone had taken their turn and I was the only one left, I broke the board with one solid chop. I took the pieces to my office and placed them in my wall locker—I wasn't ready to discard them yet. From time to time I'd see or glance at my broken board, but it stayed in the locker until the day I packed up my office. I finally saw them as junk and threw them away.

Today, I'm able to look that fear square in the eye and say, "No matter what you think, I'm going for it." For me, not letting fear control me is more significant than letting it go.

For two decades of military service, I had put off taking college courses because I didn't think I was smart enough to pass the classes. After much trepidation—and determination to *walk the talk* with my Airmen—it was time to enroll. After the first class, I thought, "What have I been so afraid of? I got this!" I overcame my fear and was able to take the rest of the classes—even Intermediate Algebra. After retirement, I continued to stare fear straight in the eye and pressed forward to complete my bachelor's degree.

There is no time for fear when you're busy taking action! Never let the fear of failure stop you from trying something new or achieving your goals. The only failure is to stop trying. ∎

Lesson Three: Moving On after Mistakes

The only man who never makes a mistake is the man who never does anything.
~ Theodore Roosevelt

S lip-ups, blunders, errors in judgment, and failures are inescapable and universal; everyone makes them. But what a person does *after* something goes wrong defines who they are as a person. Do you beat yourself up after making a mistake, or do you use them as opportunities to gain wisdom? Most leaders will tell you they learn much more from their failures than from their successes.

1. Give yourself the grace to fail

You are your own harshest critic. Having the grace to accept and understand that you're not perfect allows you to move forward and do better the next time.

2. When mistakes happen, let them go

Life is full of situations where "stuff happens," and you have to be able to adjust on the fly, find the positive, and move forward.

After three years of serving as the 154th Wing CCM, the Hawai'i CCM position became available—the highest enlisted position in the Hawai'i Air National Guard. True to form, I over prepared. I rehearsed my opening and closing remarks in front of the mirror and practiced appropriate hand gestures. I anticipated questions and developed responses. I meticulously went over my service dress uniform and re-measured the placement of my U.S. insignia, ribbons, occupational badges, and nametag.

The day of the interview, I hand-carried my uniform to the Hawai'i state headquarters so it wouldn't wrinkle. As I changed, I caught the corner of my magnetic ribbon rack and it fell to the floor. At the time, my ribbon rack was a perfect square. No problem—I quickly placed the ribbons back on my uniform, aligned them, and walked to the conference room.

During the interview, I had an uneasy sense that something was off, even though I was answering everything well. Several times, I noticed Maj. Gen. Robert Lee, the Hawai'i adjutant general (TAG), looking at my ribbon rack. I felt uncomfortable, but pressed ahead. After the interview, I caught a glimpse of myself in the mirror. To my horror, I saw that my ribbon rack was upside down. I was mortified. It was a huge faux pas, especially for someone so detail-oriented. My first thought was to go back in to the board and apologize, but I decided that would just make it worse. I left the building thinking I would not be selected.

When I received a call from the TAG later that day, I braced myself for bad news. I was genuinely shocked when he said he'd selected me to be the Hawai'i Air National Guard State Command CCM. But even after he'd congratulated me, I couldn't let go of the ribbon debacle. I needed to know if he'd noticed and whether or not that had become a point of discussion in the selection process.

I realized I could continue to beat myself up, but I prayed that God would take it from me so I could let it go. I had done all the right things to prepare, and *stuff happened*—such is life. I continued to work hard and reminded myself that everyone makes mistakes. Years later, I learned that Major General Lee had not even noticed my upside-down ribbon rack! I thought about all the precious time I had wasted worrying about something I could not change.

3. Find a lesson

We often don't take adequate time to really think about our mistakes or failures. Ask yourself, "What could I have done differently? What did I learn from that situation? What can I do to avoid future missteps?"

As I took great pride in always having my uniform squared away, this experience truly humbled me. I finally realized the silver lining to this embarrassing incident: I had a new perspective when I saw an Airman or Soldier who needed a uniform correction. I had always approached them with dignity and respect when making corrections, but now I had more compassion and understanding for others' missteps and oversights. Sometimes that might be the only lesson you gain from a blunder, but it's an important one. Uphold the highest level of standards, ensure compliance, and be a strong role model, but realize we're all human. If you make a mistake, commit to doing better next time and then let it go. ■

Lesson Four: Positive Attitude

Nothing can stop the man with the right mental attitude from achieving his goal;
nothing on earth can help the man with the wrong mental attitude.
~ Thomas Jefferson

G rowing up, I didn't understand the concept of choosing your attitude. When you're a kid, you think your emotions are out of your control—you just can't help it if someone makes you mad or hurts your feelings. Frankly, we weren't allowed to express much emotion growing up on a farm, and for years I learned to stuff my feelings down. It wasn't until after basic military training that I realized only *I* can choose my attitude.

Halfway through ATC school, I received notice that after graduation I would be assigned to Laughlin Air Force Base in Texas. I was elated! This Minnesota gal was not going to be cold or have to shovel snow for at least the next three years. A few days later, while basking in the glory and thoughts of warmth and sunshine, I received a new set of orders diverting me to Offutt Air Force Base in Nebraska, where the weather was just as bleak in the winter as Minnesota's.

What?! I thought, devastated. How could they send me back to the tundra? What happened to the pitch, "Join the Air Force, travel, and see the world"? Although I didn't show my disappointment, inwardly my attitude plummeted. On the flight to Offutt AFB, I made a conscious decision to change my attitude. I decided I was going to be the best Airman I could be. I would train hard, learn the material, volunteer, and do everything the United States Air Force asked me to do.

Abraham Lincoln said, "Most folks are about as happy as they make up their minds to be." I discovered this to be true. Changing my outlook made the difference from being a mediocre Airman to being a top-performing Airman. Attitude is a mindset we can control. Every day we need to decide to bring our very best self and very best attitude to the work center, to the fight, and back home at night.

During times of adversity, I looked for the silver lining. My rule was that I would give myself three days to stew on something, and then I let it go. Typically, in that time I found the reason or purpose for the situation. Sometimes it took longer than three days before I found it, but three days is all I'd give to disappointment. During this time, I never let it affect my attitude and I remained positive and upbeat.

The first time I applied for the Hawai'i Air National Guard state CCM position, I wasn't selected. It hurt, and it really knocked the wind out of my sails. In 19 years of service, I had only once finished second in a board selection. I took my three days to stew and mull it over, but I couldn't find the silver lining in not being selected. Nevertheless, I kept doing my best and never quit believing in my abilities. I continued to be a good Airman, and I didn't let the experience get me down.

Months later when I was selected for the position of the 154th Wing CCM, the silver lining finally became crystal clear in my mind: I was meant to serve at the wing level. I needed to get to know the officers and enlisted leadership and acquire a deeper understanding of the wing's multiple mission sets. I realized that by maintaining the right attitude and being mission-focused, rather than feeling rejected and downtrodden, made a significant difference in how I viewed the non-selection at the state level. And, as it turned out, the wing CCM job turned out to be the best thing that could have ever happened for my career and for me personally.

Dr. John Maxwell, one of my most influential mentors in leadership, paraphrases Zig Ziglar: "Your attitude will determine your altitude." Your outlook on a situation makes all the difference. The best thing about attitude is that we have complete control over it. Choose your attitude carefully, as it will determine what you do with your life.

A good attitude creates the tipping point: all things being equal, the one with the best attitude will win or get ahead. Carrying a good attitude into all I did was critical. I didn't let the naysayers get me down. Generally speaking, supervisors tend to select those people in their organizations who have good attitudes and are positive, so I made sure I had the right temperament. This isn't to say I never got down or wasn't disappointed—I just didn't allow myself to *stay* down. After those three days of stewing, I always made a deliberate decision to move forward with a good attitude. Dr. Maxwell calls it "failing forward." I have lived by this principle repeatedly in my professional and personal life. We all have challenges, struggles, and difficult times—that's life! There's no way around it. But the key to a happy life is to look for the positive even in the worst of times. This outlook will help you get through many tough times. ■

Lesson Five: Be Ready

To each there comes in their lifetime a special moment when they are
figuratively tapped on the shoulder and offered the chance to do a very special thing,
unique to them and fitted to their talents. What a tragedy if that moment finds them
unprepared or unqualified for that which could have been their finest hour.

~ Unknown

This quote[2] on the value of preparation has been one of the cornerstones of my life. No one knows exactly when he or she will be tapped on the shoulder to lead, assume higher levels of responsibility, take on a new challenge, or be offered an incredible opportunity. To be ready for that moment, we need to look ahead, plan, and be open to opportunities. Long before that time comes, we must find our reason for being, identify our gifts and talents, and put them to the best use possible.

I was not ready for my first leadership opportunity at Basic Military Training (BMT). As dorm chief, one of my responsibilities was assisting other trainees with preparing for their wall locker and drawers inspection. I was expected to take the lead and show my fellow trainees what constituted perfection and how they needed to prepare. Locker inspection day, however, came faster than I had anticipated. I quickly spaced my hangers in my locker, aligned my clothes in my drawer, and tightened my bed corners. My sister trainees and I waited for the inspection results.

When I heard the TI call my name, I thought, *This can't be good.* Staff Sergeant Earnest pulled me aside and quietly told me that he could have redlined me because I had excess threads and quality inspection stickers on my uniforms. I had no excuse. Helping others at the cost of not preparing my own items was indefensible. However, he said he wouldn't fail me because I was the only thing holding the flight together. I thanked him and said my locker would be squared away that evening. I respected how he treated me, and his example of correcting in private and praising in public stuck with me. From that point on, I was ready for an inspection at any time of the night or day. I didn't wait for an official inspection—I was prepared all the time.

Readiness in all things is engrained in military members from the very beginning of their military service. That doesn't just mean being prepared for an emergency or for a deployment with 48 hours' notice. Readiness encompasses not only being prepared to serve in your civilian and military career, but also in your family life. Military members understand that when "the call" comes, the whole person must be ready. Relationships and family life must be strong. You must have a family care plan, with medical, dental, and physical fitness current, and make sure your mental and emotional health is solid.

---

[2] Although the quote is popularly attributed to Winston Churchill, no verified source shows that he said it or wrote it. However, in *The Gathering Storm*, Churchill's first volume of his history of World War II, he wrote something similar in meaning: "I felt as if I were walking with destiny, and that all my past life had been but a preparation for this hour and this trial." Winston S. Churchill, *The Gathering Storm* (Cambridge, Mass.: Houghton Mifflin Co., Riverside Press, 1948).

Are your finances in order? Are your vehicles in good repair? All these things and more are essential to readiness—to drop everything at a moment's notice to be available for a new challenge, or to be mentally and physically equipped when you are offered a chance to shine.

Ask yourself, "What do I need to do to be ready for the next leadership opportunity?" Think about what distinguishes you from your colleagues. Unlike an opening for a promotion, leadership opportunities are those occasions that are not necessarily obvious. Sometimes you have to look for them and seek them out.

Readiness means that learning and growing consistently needs to be at the forefront of what we do. Whether you're in the military or a civilian, annual and recurring technical training and higher education is paramount. As the command chief for the 154th Wing, Hawai'i ANG, I saw all too often opportunities pass by Airmen because they hadn't completed their professional military education (PME). My advice is to enroll for and complete your next level of PME as soon as you are eligible. Completing each level is essential not just for promotion; more importantly, it prepares you to assume higher levels of leadership. Continual professional and personal education gives you additional tools and resources and enables you to think more broadly. Promotions will come when you can check off the right boxes, but having extra tools in your rucksack will help you grow and be a better leader.

While serving in the Pentagon, I interviewed a chief master sergeant for a highly sought-after position. Unfortunately, he lost out on the opportunity because he had not completed his two-year degree from the Community College of the Air Force (CCAF). He had one class left: math. (I understood his apprehension!) The time to prepare is not when the opportunity appears.

You can prepare to be a leader in many ways. When I was a young Airman doing air traffic control training in Hawai'i, it was extremely difficult to attend PME courses in residence. Gary was an active-duty Marine who was often gone off-island for temporary assigned duty, and our daughter Ashley was very young. PME in-residence courses would have taken me away for four to six weeks at a time, and we had no family or friends in Hawai'i who could keep Ashley for that long.

I completed my PME by correspondence, but it was done in isolation. I had no opportunity for interaction between instructors and other Airmen; I didn't have the chance to ask questions, give presentations, or participate in discussions. I needed to find a way to make up for the disadvantage of not attending courses in residence, so I took the initiative to pursue some education on my own.

I began to read books and articles on leadership and management. I made my driving and running time productive by listening to tapes and CDs. (Today, podcasts are readily available and usually free.) Many days I listened to half of a leadership lesson on the way to work and the other half on the way home. I applied the leadership tips I'd gained in my interactions with those for whom I was responsible.

One of my favorite leadership programs is Dr. John Maxwell's Maximum Impact Club. I received a monthly CD that I listened to several times so it would sink in. I was inspired by many of Dr. Maxwell's lessons and created several speeches based on the information I learned. Taking the initiative to develop my talents and abilities prepared me for leadership opportunities when they came.

My advice to anyone looking to advance their career is to develop skills outside of PME to make yourself more competitive. Learn and understand the "tipping points"—all else being equal, what determines that you are selected for an opportunity instead of someone else—and employ the tactics necessary to set yourself apart from your peers. Engage early and often and do a pulse check on where you are on your journey. Do everything you can to be ready. Don't leave your career—or your life—to chance. ■

Lesson Six: Take Care of Your People

Leadership is a two-way street: loyalty up and loyalty down.
Respect for one's superiors; care for one's crew.
~ Rear Adm. Grace Hopper

The U.S. military is the best in the world. We have the latest technology, unparalleled military training, and state-of-the-art equipment, weapons, aircraft, tanks, and artillery. Our most important asset, however, the one that sets us apart from the rest of the world, is our people. We need to take care of the men and women who complete the difficult tasks and get the mission done.

"Taking care of your people" means eyes-on leadership. It means unplugging, walking the halls, the flight line, and the back shops. It means interacting: asking questions, listening, observing body language, and then asking deeper questions.

My third TI in basic training was that kind of a leader. Sgt. D.R. Fox was four-feet-something and had just returned from maternity leave. She trusted my judgment as a dorm chief and asked me to write her a nightly report. This was long before computers, so I wrote the reports by hand and usually finished around midnight. I spent so much time writing these detailed reports, overseeing the flight, and attending to other responsibilities that I hadn't had time to study for the upcoming test on the basic training manual. I was starting to feel anxious and overwhelmed.

Late one night, Sergeant Fox called up to the flight through the speaker to check in. Even over the intercom, she picked up on the emotion in my voice, and came up to see me. When I explained that I hadn't had time to study, she immediately took action to help. She chose one of the smartest trainees in our flight to help me learn the most relevant material, and I passed the exam. Sergeant Fox showed me how a good leader behaves: she was sensitive and attuned to a change in my voice that showed I was not my usual self. She listened to my concerns and then took swift decisive action on my behalf.

Some leaders say, "I can't ask my people about their personal lives; it's not my business." Oh, but it *is* your business! Listening to and observing your people—truly getting to know them and understand their needs—is critical to the successful completion of the mission. It's your job to know if there's something troubling going on in their life. Make eye contact, notice body language, and ask questions. Taking the time for one-on-one leadership could very well save someone's life.

Some years ago, I was the guest speaker for a noncommissioned officer academy (NCOA) graduation in Knoxville, Tennessee. As I watched the flights lined up on stage to be recognized, I took note of one particular female NCO. She looked like she was carrying the weight of the world on her shoulders; her head remained down and a look of despair was on her face. What should have been one of the happiest days of her Air National Guard career appeared to be one of significant and deep duress. I knew something was definitely wrong.

After the graduation, I tried to find her among the hundreds of Airmen all dressed alike. I believe it was through divine intervention that she happened to walk by me. I approached her and asked if she was OK. Clearly distressed, Staff Sgt. Patty Winner told me that her ex-husband had called her that morning to say he was taking their daughter and filing for sole custody.

She was far from home and completely at a loss about what to do. I got her cell number and told her I would contact her state command chief so she would get assistance. I also spoke with her flight leader at the academy and asked him to check on her before she left for home. Later that week, I followed up with Staff Sergeant Winner to make sure she was okay. Her state command chief had coordinated with her wing command chief and first sergeant, who connected her with the wing Family Readiness Program manager and the director of psychological health. They assisted her with resources to deal with the legal and psychological situations.

A few years later, after I had retired, I was delighted to see Staff Sergeant Winner at an event in Washington, D.C. I didn't recognize her at first—she had a bright, big smile on her face, and looked and sounded happy and confident. She told me she had custody of her daughter, a new job, and had been promoted to technical sergeant. It was such a blessing to see how she had changed and how strong and resilient she had become.

I asked Technical Sergeant Winner—now a master sergeant—if I could share her story, and received a deeply meaningful letter in response. I'm including excerpts here to show how leaders can make a huge difference by taking a personal interest in their people:

> The timing for the NCO academy could not have come at a worse time. Mentally, I felt prepared for the training. Emotionally, I was scattered, not stable, just not myself. My confidence was low, and participation and interaction with others was gradual. I encountered sleepless nights worrying about home. I was devastated, but tried to stay focused on training.
>
> Come graduation day, I was filled with anxiety as I anticipated the challenges to come. As the ceremony finished up, family members, unit leadership walked around congratulating their Airmen and loved ones. I was just about ready to depart the building as you approached me. You grasped both my hands and congratulated me, introduced yourself and said you watched me walk up with my class and receive my graduation certificate. You saw within me something that concerned you. You said, "I want you to know that whatever you are going through right now, I am here for you. I will be your wingman and we will get through this."
>
> I drove home to Pennsylvania, ten hours, thinking about what you said, wondering how in the world you could see in my expression the challenge and despair in my life at that time. Our paths crossed at a very specific and critical time in my life. Your compassion gave me strength and the drive to get back to my civilian life and take on the challenges to come. Your actions are a proven example of a wingman and have provided me with a true model of professionalism to live by.
>
> In many ways, our encounter was a turning point for me. Sometimes all it takes is someone telling you exactly how they see it, knowing that they believe

in you and hearing someone else say we will get through this, because for some time I was contemplating, "Will I get through this?"

When I'm overly stressed, I try my best to seek out friends, some type of support. But life gets in the way, and before I know it, I'm on the verge of crumbling. A couple weeks ago, when I got your email, it came to me as a reminder that someone cares and that I can get through this.

Ma'am, you've instilled within me the hope and desire to be strong and provide others strength when they are faced with challenges, be it embarking on a new journey or facing a hurdle to the finish line. There is a leader within me. I know that now. In the simplest of phrases that could possibly capture the magnitude of your kind gestures, you saved my life.

My sincerest gratitude,
TSgt Patty Winner

Everything we're taught as Airmen and leaders about the importance of caring for our people is evident throughout her story. Make the time to know your people! You never know how your interaction and concern will help someone.

To get the best out of your men, they must feel that you are their real leader
and must know that they can depend upon you.
~ Gen. John J. Pershing

Lesson Seven: Recognize Accomplishments

Appreciation can make a day, even change a life.
Your willingness to put it into words is all that is necessary.
~ Margaret Cousins

Taking care of your people includes routinely reviewing their records, giving feedback and course corrections, and recognizing them for excellence. In my experience, recognition in the form of decorations is sparse in the National Guard, because members belong to a specific state during peacetime and don't rotate or move every few years like our active-duty counterparts.[3] Some states and units do a phenomenal job recognizing their people, but many do not.

Recognition was not important to me throughout most of my career; my service was not motivated by an award or medal. Later in my career, however, some people made mistaken judgments about me based on my lack of decorations. At my wing CCM promotion ceremony, I met Capt. David Lowery,[4] aide to the commander of the 15th Air Base Wing (ABW). We chatted for a little while and then he looked down at my ribbon rack. I was taken aback when he said he had never seen a chief, let alone a senior master sergeant (which I was at the time), get promoted to CCM while only having an Air Force Achievement Medal with one device as their highest decoration.[5]

It stung a bit to hear Captain Lowery's comment. Unfortunately, my lack of decorations was pretty typical: many Airmen didn't get recognized for doing their jobs, their decades of service, or even when they performed exceptionally well. Overall, HIANG Airmen were not appropriately acknowledged for their dedicated service, achievements, and significant contributions to the federal and state mission.

Not too long after that, I was at a command chief conference when a fellow active-duty CCM asked, "How did you screw up and still manage to become a command chief?" I was taken aback and didn't understand what he meant until he pointed to my ribbon rack. I assured him I was *not* a screw-up but that I had belonged to an organization that did a poor job recognizing its people with decorations. I explained that as a new wing CCM, I was committed to changing that culture. That was the first time I had ever felt embarrassed because of my lack of ribbons.

A few months later at another conference, an Air National Guard CCM asked me about my lack of decorations. As a Guardsman, he understood the culture in some organizations and states. We had a lengthy discussion about the situation and, once again,

---

[3] When active-duty personnel transfer to another base or duty station, they often receive an "end of tour award" for their service, significant accomplishments, performance, and contributions during their tour of duty. Members of the National Guard and Reserves don't transfer like the active services and thus do not receive that award, so their performance and contributions are routinely overlooked or taken for granted.
[4] Now Lt. Col. David Lowery, Ph.D.
[5] Airmen receive their first decoration in the form of a ribbon, and subsequent awards are presented as a device, an oak leaf cluster placed on or affixed to the original ribbon. I had one ribbon and one cluster.

I left feeling less than equal to my counterparts and colleagues—and a bit embarrassed. Overall, most Guard CCMs, and certainly all active-duty CCMs, had an appropriate level of decorations for their rank, contributions, and time in service. I was not one to write a personal decoration, although some have been asked to do so. I believed that if leadership deemed me worthy of a decoration, they would write and submit a package.

After those comments, I began to take note of others' ribbons and decorations, and I became very self-conscious about my sparse ribbon rack. Active-duty chiefs had significantly more and higher-level decorations than many of their Guard counterparts, but I understood this was mostly due to a well-deserved end-of-tour award. Decorations do tell your story. A *lack* of decorations can tell yet another story, which may not be accurate. Some may believe a lack of decorations means poor performance, mediocrity, or a troubled past.

In the Hawai'i ANG culture during my service, there was no expectation of receiving awards or decorations. I came to realize I'd probably accepted a lack of recognition because that's the way I had grown up. In my family, you did what you were supposed to do and there was little fanfare or acknowledgment for doing it.

I'd left the 297th Air Traffic Control Squadron (ATCS) as a senior master sergeant, having been the ATC superintendent, squadron, and group NCO of the Year; Squadron, 201st Group, and HIANG Air Traffic Controller of the Year; and later the HIANG SNCO of the Year, recognized by the Inspector General during the 241st ATCS Operational Readiness Exercise and privately during our "Excellent" Operational Readiness Inspection. I had also received two Air Force Achievement Medals, one from my active-duty service and one from the Air National Guard Readiness Center for support provided. During the 12 years in the 297th ATCS, I received longevity devices and a state ribbon for deploying to Hurricane Iniki in 1992 on the island of Kauai. Whether or not I earned or deserved other decorations was not for me to say; it was up to my supervisor and leadership to make that determination, and that simply was not done in the 297th.

I was grateful that Captain Lowery offered to look at our decorations procedures and collaborate with me to fix our awards program. I quickly learned that countless deserving Airmen had not been recognized for many years, decades even—some never had been. It was not unusual to see a master sergeant with 14-plus years who had never received an Air Force Achievement Medal. Packages were submitted but not properly tracked, and many were lost. Airmen referred to the awards and decorations program as a "black hole": things went in but never came out. Many felt it was a waste of time to submit a package. As I went from unit to unit and attended meetings, this was the number-one piece of feedback I received from Airmen: fix the awards program!

The bigger picture was even more important. While leadership was slow to recognize exceptionalism, we were often quick to react to poor performance. Recognition—or the lack of it—affected both individuals and organizational morale.

We recruited a group of subject-matter experts and established an awards and decorations action team. Over the course of several months, we rewrote the HIANG Awards and Decoration Instruction. The process was streamlined and simplified, and the approval authority was delegated at an appropriate level. We had committed Airmen who dedicated time to ensure the program's success.

The next hurdle was to convince officers and enlisted members to reconstruct and resubmit packages retroactively for the many years of service and performance that had gone unrecognized. This took a great deal of convincing and there was enormous skepticism. After nearly a year of hard work, our Airmen began to be recognized for the exceptional work they had previously done and were currently doing. It was not long after getting this program up and running that the U.S. Air Force changed the entire awards program for all components (Active, Guard, and Reserve), which made it even simpler to submit an awards package.

I found it humbling that it was an active-duty captain who took the time to write an awards package for my Air Force Commendation Medal, after 21 years of service. I had then been a command chief for 19 months. Shortly after receiving the Air Force Commendation Medal, I received my first Meritorious Service Medal for my 12 years of dedicated service and my contributions with the 297th ATCS.

Throughout this process, I came to understand why it is important for leaders to routinely review service records and recommend meaningful awards and decorations. Service members sacrifice, deploy, volunteer, and do extraordinary things every day. Being recognized for a job well done lets people know their contribution matters. What we wear on our uniforms or display in our offices tells a story of service, sacrifice, and personal and unit accomplishments. Some may say that recognition and awards aren't important to them, but leaders need to take the time to do it anyway for those who deserve it. Later on in their careers, it *will* matter. Don't allow others to judge your people because you didn't do your part—appropriate recognition is a crucial element of taking care of your people. ∎

Lesson Eight: Expectations and Accountability

Accountability breeds response-ability.
~ Stephen R. Covey

The vast majority of service members do an amazing job of maintaining standards and their professionalism. Some, however, relax their standards, overlook minor infractions, and let personal desires cloud their judgment. Over the years, I've learned that bad behavior often results from a power struggle, lack of self-confidence, low self-esteem, or because the individual was never reprimanded. Leaders at all levels have an inherent responsibility to never tolerate bad behavior and to hold accountable those who violate the rules.

In my opinion, an unwritten rule is that supervisors and senior leaders should never drink alcoholic beverages with their subordinates, other than in a social setting like a banquet. Drinking alcohol in excess reduces inhibitions and compromises professionalism. Standards and lines of professional decorum exist for a reason. Alcohol can make it far too easy to cross those lines—with detrimental effects.

When I was a senior master sergeant, I once volunteered at a conference with several subordinates from my unit. One evening event consisted of line dancing, mechanical bull riding, and other Western-type activities. Hundreds of Airmen from throughout the ANG attended. I had one glass of wine as I walked through the crowd and participated in the line dancing. I enjoyed myself and visited with many people, then turned in early.

A few days later, I was talking with our group commander about the event. A technical sergeant overheard our conversation and said, "I saw you drinking and partying." I was aghast at her remark and her perception of my behavior. I quickly explained to the commander that I was not "partying" and had only had one drink. From that day on, I was overly cautious about consuming any alcohol where junior enlisted were present, other than at an event like a formal ball.

I also learned how important it is to set a drink down before having a picture taken. With social media, you can never retrieve inappropriate pictures, and you never know when they will come back to haunt you or have an impact on your career. Perception is always someone's reality.

I believe professionalism extends beyond the office or the end of the official duty day. The American public holds our military to a higher standard of professional and personal decorum. This is something communities across the U.S. and around the world have come to expect from our nation's finest—and we must deliver. Someone will always be looking and making judgments based on your behavior, so live as if you're in a glass fishbowl 24/7.

Leaders must enforce compliance with standards and accountability by always setting the example. When I served in the ATC Squadron as the superintendent, the longtime standard was that on drill weekends every controller needed to complete and document their recurring, monthly, and annual training in their training record. Although this standard was well known and clearly explained, the documentation was haphazard and lax. One drill weekend, fed up by this lack of compliance, I told the controllers that no one was authorized to leave until the training was done and documented and I had reviewed the training records. I was amazed at how quickly the training was accomplished, signed off, and brought to me for review.

Soon after becoming wing CCM, I met with the first sergeant team. I noticed the security forces first sergeant was wearing a beret. As a first sergeant, he was not authorized to do that—he should have been wearing the issued battle dress uniform (BDU) hat. Together we read the uniform standard and had a long conversation, but he refused to comply because the previous CCM had allowed him to wear the beret.

"I'm the new command chief, and I'm telling you you're not authorized to wear a beret," I said firmly. "You have three choices: take off the beret and wear the BDU hat, transfer back to your old position, or retire."

While I understand the importance of the security forces defenders earning and wearing the coveted beret, I had to enforce the standard. He chose to retire.

During my time as the 154th Wing CCM, the Air Force introduced the Air Force Physical Training Uniform (AFPTU). Initially, we had challenges getting all Airmen to comply with the standards. Senior leaders and first sergeants spent time explaining the standards and educating the entire force. With primarily a traditional drill-status force, it took additional time to enforce changes and to get the word up and down the chain to every Airman. As the functional manager of the first sergeants, I held them accountable to ensure the standards were being met.

After I transitioned to the state command chief position, I met with the new wing CCM, Chief Master Sgt. Robert Lee III, and we discussed how we could help get the word out and better enforce the AFPTU standards. We decided to engage the wing's multimedia department and make poster boards showing the two of us wearing the PT uniform per Air Force instruction. We strategically placed these poster boards in high-traffic areas like the Medical Group, dining facility, and the Force Support Squadron. It was encouraging to watch Airmen correcting each other as they stood in line for medical appointments or lunch. Mission accomplished!

Anyone who knows me knows I'm a stickler for standards. I would not let infractions go by without providing professional corrective actions. I believe in accountability and was not intimidated by rank—even the highest ranks are not immune to dress and appearance corrections. It's all in the approach that makes the difference. I've seen some leaders make corrections by yelling or publicly embarrassing people. That is not how corrections should be made. The person being corrected needs to learn something. Often I'd ask them in private if they knew what the standard was, and many times I'd ask them to look it up and get back to me.

My experience has been that whenever there was a uniform change, it took time to get everyone in compliance. For a brief time, the Air Force allowed black pearl earrings. In Hawai'i, black pearl earrings are very popular. Once that change was rescinded, it took months to make all the necessary corrections, educate the force, and ensure compliance.

One day when I was a drill-status Guardsman (part-time), I left my civilian job and went to the control tower to perform ATC duty. I changed into my uniform but forgot to take off my earrings. No one said a word to me and it wasn't until after my shift that I caught my blunder. The next day, I asked the crew why no one had said anything to me. A couple of them had noticed but were too embarrassed to correct me. I encouraged them to let me know anytime they saw me with a uniform discrepancy. I would have welcomed the correction and complied immediately.

While walking the halls in the Pentagon, I often saw uniform violations. If I had time for a short conversation, I would stop and politely discuss the infraction. One day, I saw a female major wearing hoop earrings while in uniform. I asked for a moment, and privately told her she may not be aware she had earrings on. She was appreciative and immediately took them off. Others have not been so receptive. I once corrected an Airman for not wearing a hat/cover while walking through the parking lot outside.

"Oh, sorry, I get it," she said, smiling. "When the chief's around I have to wear my hat."

"No, that's not it," I responded. "The rule is you must wear your hat at all times while outdoors! It has nothing to do with me being around."■

Lesson Nine: A Man's World

*How wrong it is for a woman to expect the man
to build the world she wants, rather than to create it herself.*
~ Anais Nin

I don't have a simple answer when people ask me what it was like to be a woman in a man's world, as the military was during much of my career. First, I explain that when I was growing up in the traditional '60s and '70s, my father ruled the house. He ran a tight ship, and there was no questioning his decisions or talking back. He barked orders; we executed them. Thus, it was no surprise that I chose the Air Force, a male-dominated career field, or why I felt comfortable under the leadership of male supervisors. I was used to it.

I was driven to succeed and did all I could to ensure I learned my job and how to be a good Airman. I never felt limited because I was a woman. After transferring from active duty into the California Air National Guard, I not only did what my supervisors asked of me, I went above and beyond the call of duty and maintained high standards. The Ground Radio Operations career field I cross-trained into was fairly male-dominant, but I was used to having a man provide training and direction.

It wasn't until I was an NCO at about the technical sergeant rank that I began to see a glimpse of what some people called "the good ol' boys' club." Nevertheless, I still didn't buy into the notion that "it's a man's world" or that "only men can get ahead." I believed that one's record should speak for itself, and I continued to be the best Airmen I could be. My uniform was always squared away. I completed my PME ahead of schedule, volunteered, and maintained a positive attitude. I had to stand up for myself from time to time, but some men have to do that as well.

As an SNCO, when it came time for promotion or for assignment into a new position of leadership, I did *not* let the fact that I was a woman enter into the equation or become part of the conversation. Even on Oahu where I was a *haole* (non-native Hawaiian) and didn't look like the majority of the Airmen, it just didn't matter. I refused to play the "woman card" or allow it to be played on me. I believe if a woman does her job, completes all her required training, volunteers, and is a well-rounded Airman, her record stands on its own merit.

A few of my key recommendations for female warriors in today's military:

1. Find female mentors

One day while assigned to the Headquarters 201st Combat Communications Group as a full-time technician, I attended an off-site senior leadership day of training. Colonel James Townsend was leading the discussion on mentorship. He was going around the room, asking who our mentor was and why.

My mind raced to come up with a name.

Mentorship had not been part of my journey thus far. There had never been a supervisor or leader who had taken me by the stripes and provided guidance, recommendations, or exposed me to new opportunities. What does one do in a case like that? You figure it out on your own! I had been doing that for most of my career—and in my life, for that matter. Although I looked up to my parents and grandparents, they had never mentored or counseled me. As a teenager, I had learned to make decisions on my own, based on the values they had taught me.

"Chief, who is your mentor?" Colonel Townsend asked.

I exhaled with a deep sigh. "I don't have one," I admitted. "I just didn't grow up in that kind of environment. But . . . there is someone I've looked up to and admired since I was a teenager: Mary Tyler Moore."

That got a few laughs and chuckles.

"Let me explain!" I insisted, holding up my hand. "Mary Tyler Moore's TV character worked in an all-male newsroom, and she had to work harder than her peers and prove herself every day. She did this with grace, tenacity, and great poise, and I believed that if she could do it, so could I. Mary never gave up and never quit, and that was something that has been instilled in me since childhood."

2. Accept and understand challenges

Being a woman in an all-male or mostly male environment sometimes did have its challenges: for example, to make my voice heard or to make a point in a meeting, I often had to interrupt the conversation. Female voices are simply not as loud as male voices, generally speaking. But I found that as long as I didn't mind getting my hands dirty, I fit in with the guys. I was not afraid to roll up my sleeves and work hard. If we were in the field, I had no problem sleeping in the tent with dirt floors or being in MOPP4 gear (a full chemical training suit). I actually enjoyed the combat field condition environment.

3. Don't be afraid to stand out

While assigned to the Pentagon, I directed my staff to wear our service or class "B" uniform in the office. However, when I visited troops in the field or during a Unit Training Assembly weekend, I wanted to mirror their uniforms, which was usually the Airman Battle Uniform (ABU). I made a point to wear my jacket and skirt when the service senior enlisted advisors (SEAs) and I were together in service dress uniform. I did so because while I represented all Soldiers and Airmen in the National Guard, I also felt a strong sense of obligation to the female warriors, many of whom did not know their National Guard Bureau SEA was a woman. I wore a skirt so I wouldn't blend in with my male counterparts. I wanted female service members to see and understand that they, too, could achieve the highest enlisted position in their respective branch or component of service. This was not something I specifically talked to women about, but the subtlety of the skirt helped make my point.

Meetings in the Pentagon were respectful; if you had something to say, you merely had to raise your hand. From the President, Secretary of Defense, Chairman of the Joint Chiefs of Staff, service secretaries, and service chiefs to senior military leaders and civilian officials— they all wanted to hear from the SEAs. I never felt like I was shut out, stifled, or ignored.

During my interview for the position of CMSAF, I was asked, "Chief, if you're selected, you will be the first woman in this position. How would you address this unique circumstance?" My answer came without hesitation.

"Our uniform is gender neutral, Sir," I said. "If selected as the CMSAF, I would have the responsibility to represent all Airmen, regardless of gender, and to be the best role model for them. I believe in standards—not separate male and female standards, but THE standard. I've always believed the best person for the job should get the position. Of course, if I were selected as the first female CMSAF, I would be honored to represent our women warriors."

4. Don't tolerate disrespect

Occasionally, I needed to correct men who spoke disrespectfully. As a master sergeant, I took a 30-day tour at the Air National Guard Readiness Center in the ATC Directorate. It was an amazing learning experience. I was able to bring back administrative ideas and examples of how to improve our operations to our unit. Initially, my renewed sense of purpose, passion, and energy was not well received by the Title-5 (civilian) training manager, which became obvious during the commander's staff meeting. As I was explaining what needed to be included in a certain report, the training manager became highly disrespectful and sarcastically said, "Then tell us what needs to go in it, Ms. Smarty Pants."

I looked at him straight in the eyes and didn't flinch.

"Don't you ever speak to me like that again," I responded firmly.

The commander shut down the dialogue and moved on. The training manager and I later had heated words and resolved the conflict. I don't know why he thought it was OK to speak to me like that—I'm sure he would never have addressed a male counterpart that way.

A similar situation happened while participating in an exercise in Korea. I was a chief master sergeant at the time and known to be early to meetings or appointments. As I walked to a meeting, I got lost and was about ten minutes late. When I entered the room, the civilian presenter took note of my tardiness in front of the entire room. Then he said, "You're the youngest chief I've ever seen! I wonder what you did to get promoted." I was appalled and insulted. I remained standing while I fixed him with a steely gaze.

"I can assure you I worked hard for it," I said calmly.

After taking my seat, I immediately recalled the earlier incident with the training manager and again wondered why some men thought it was OK to speak to me like that. I don't have a "victim" mentality about being a woman, but I do wonder if any of my male peers ever had to deal with this type of unprofessionalism. My guess is… no.

To use a baseball analogy, I often felt like my starting line was a little behind my male counterparts. The men hit a home run, but they started on third base—while I had to run all the bases. Never one to quit or give up, I ran the bases and most of the time made it to home plate. I was raised to have a thick skin and not be easily offended. I didn't let these incidents faze me. However, not all women can brush hurtful behavior off, nor should they have to. Male or female, no one should have to listen to unprofessional remarks. If disrespectful speakers are uncorrected, others may believe they too can talk to women that way.

5. Focus on your willingness to serve

Ultimately, though, I don't believe it mattered that I was a woman, in terms of getting the positions I held in the Air Force and National Guard. I've often said that it's important to serve "pure of heart," meaning that if you serve for the good of the country and your organization, and if you do what's right for the people you lead, great things will come back to you every single time. I believe the "right person" for the job depends on your willingness to serve and your record, not your gender.

6. Hold fast to one standard

I was often asked about standards. Without exception, female service members did not want lowered or separate standards. I'm a firm believer in one standard. If a man or woman can meet that standard, they should have the opportunity to serve in whatever capacity they compete and qualify for—without exception.

My advice for our women warriors is to let your record speak for itself. Adhere to and maintain the highest level of standards and personal deportment, and be gender-neutral focused. Like our male counterparts, we have amazing female leaders who have and are making a difference every day.

7. Don't let evil intimidate you

Being 154th Wing CCM was the most rewarding position I'd held up to that point in my career. I had the opportunity to work with the senior leadership and develop partnerships with Pacific Air Forces (PACAF) and 15th Air Base Wing leadership. I was working with others to implement programs and policies to strengthen our enlisted corps and the overall wing. The leadership team was making positive changes, our missions were solid, and our Airmen were doing great things at the state and federal levels.

One day out of the blue, I received a phone call on my official Blackberry. The male voice on the other end was downright frightening and intentionally distorted so I wouldn't be able to recognize it. He sounded gruff, and used vulgar language to talk about raping me. I was so caught off-guard that I actually listened for a few seconds. I didn't say anything, and hung up. The call rattled me, but I figured it was random and not meant for me. Although I tried to shake it off, I couldn't stop thinking about it.

A few days later, it happened again: the same voice and the same vulgar, threatening language. This time I hung up more quickly. I knew it was no mistake: this man had my phone number and was talking about raping me. I was truly scared and didn't know what to do, but told no one. I was more careful and became more aware of my surroundings, including where I parked, but refused to let this alter who I was or what I did.

I received a third call while I was at home with my teenage daughter. It was the same vile, threatening rant, and this time I spoke. I firmly told him to never call this number again, then hung up. When she asked, I told my daughter that it was just an annoying telemarketer; then I went to the bathroom and bawled. Now I was deeply scared—it finally hit me that this man was clearly targeting me. I racked my brain, trying to think of anyone I had offended. I could not put my finger on any reason or anything I had done or said to warrant such behavior.

Perhaps this third call struck me even harder because my daughter was sitting by me. If this guy had my number, maybe he also knew who our daughter was, or where she went to school or what sports she played. Thinking about the gravity of these possibilities, I knew I couldn't stay silent any longer.

The next day, I emailed the wing and group commanders and gave them only the facts. Although shaken inside, I refused to let anyone see that side of me or let this guy get the better of me. As always, I buried my emotions and put on a good front. My husband was extra protective, but I balked and continued to put on a hardened exterior and pressed forward.

The fourth call came as a voicemail. An agent at the Office of Special Investigations (OSI) on Hickam AFB listened to it, but there was nothing they could do. Per OSI's instructions, a female Senior NCO in the Communications Flight recorded the voicemail. I was humiliated that she had to listen to it—as her command chief, it was my job to protect her from this type of vulgar language and behavior. My wing commander, Brigadier General Pawling, also listened to it and was sickened. There was nothing any of us could do, and we felt helpless.

Fortunately, that was the last phone call. I was relieved and grateful it all ended, but it took some time to put all the horrible words and emotion in a little box and place it on a shelf in the deep recesses of my memory. I had become very good at doing that. The process was all too familiar, as I'd done it since my teenage years; one more emotion or bad memory to stuff down and stow away.

I refused to show weakness or fear to anyone, especially in a male-dominant environment. Even though my rational brain told me it would have been normal for anyone to have deep feelings and a strong reaction to these calls, I did not want to be perceived as weaker than my male counterparts. Once the fear subsided, I was angry—why did women have to experience such ugly behavior?

I share this story mostly for our female civilians and women warriors. There is evil in the world, and I am deeply sorry that you have to deal with it. It's not easy to confront evil, but I know what you're made of—I know you will not stand for it. We must bring evil into the light to deal with it head-on, and we must make sure that leadership holds perpetrators accountable for their actions. We are not victims or weak. We are strong women warriors destined to compete on equal footing to be heard and to make a difference for our country.

Let the generations know that women in uniform also guaranteed their freedom. That our resolve was just as great as the brave men who stood among us. And with victory our hearts were just as full and beat just as fast—that the tears fell just as hard for those we left behind.
~ Unknown U.S. Army nurse during World War II[6]

■

---

6 This quote is etched in glass at the Women in Military Service for America Memorial just outside Arlington Cemetery.

Lesson Ten: Have Faith

I believe with all my heart that standing up for America
means standing up for the God who has blessed our land.
~ Ronald Reagan

There is a reason I saved faith for last—because I know that what I achieved would not have been possible without it. Faith in others, faith in myself, and above all, faith in God, is indispensable to me. I believe with all my heart that so many places I was privileged to see, so many wonderful people who crossed my life's path, so many things I was blessed to experience and achieve were part of God's plan for my life. I know that all things, good and bad, joyful and painful, were put in front of me to mold me into the person God wants me to be.

The path hasn't always been clear, and many times I wondered, *What is God really trying to tell me? Why is this happening to me, Lord? I don't understand!* As time went on, I began to realize that the more I quit worrying about me and the more I centered my life on others, and the more I trusted in the Lord, no matter what happened, I could be at peace with it. My life in the military and at home became a much smoother "road to travel" because I knew I wasn't alone on the journey. He was always walking by my side.

Whenever I applied for a position, I did not pray to get the job, but rather that God's will would be done. I did all I could to remain competitive, but I trusted the path the Lord had for me. When I wasn't selected, I did not fully understand why it was not my "time," but the important thing is that I never gave up, never stopped doing my best, and never quit believing in my abilities. My trust in the Lord and belief in myself continued to guide me, even in the hardest of times.

I have faced obstacles and challenges, hard times and great times. For almost 29 years I had the privilege of serving God and my country in the United States Air Force and Air National Guard. I met people from around the world, traveled to places I had never heard of, and experienced things beyond my wildest imagination. Through faith and hard work, I have achieved positions never before held by an enlisted woman. When I joined the Air Force, I had no idea that one day I would be the first woman to serve as a senior enlisted advisor to a member of the Joint Chiefs of Staff. Nor could I have ever imagined that I would achieve the highest position ever held by an enlisted woman in the history of the U.S. military. I am humbled at the life God has given me.

It is clear to me that God has always been in control of my life, and I have trusted His path. I am also eternally grateful to Gary and Ashley for their sacrifices, and for encouraging and believing in me. I could never have accomplished what I did without them by my side.

Over the course of nearly three decades serving my country, I have discovered that the military offers limitless opportunities. In the U.S. Armed Forces, you can choose to do and be anything—you just need to open your eyes and take full advantage of what lies in front of you.

Just before my retirement, I was interviewed at the Women in Service to America Memorial in Washington, D.C. At the conclusion, I was asked a very poignant question, "What will you miss the most?" Without hesitation, I answered, "A chance to make a difference." That is my reason for sharing some of my story: that you will come to know that greatness lies within each of us. My hope is that something you read will make a positive difference in your life.

Trust in the Lord with all thine heart; and lean not unto thine own understanding.
In all thy ways acknowledge him, and he shall direct thy paths.
~ Prov. 3:5-6 (KJV)

■

Lesson Eleven: Transitioning

Long-range planning does not deal with future decisions,
but with the future of present decisions.
~ Peter Drucker

L ong before retirement day, you need to make some important decisions. If you're married or have a significant other, it's a good idea to start discussing these issues with your partner at least a year before. In our household, we had two military retirements—one for my husband and one for me—so we had a lot of planning to do. The topics below are essential to consider, but the list isn't exhaustive; you may come up with a lot more on your own.

Employment
- Will you work after retirement? Full-time or part-time? What industry? What is your dream job?
- Get the certifications and education for what you want to do. This is one area you should be planning for years before retirement.
- Draft a professional résumé. You'll receive instruction on how to write a résumé in your Transition Assistance Program (TAP) class, but you can also find templates online or hire a professional résumé-writing service. Network within your circle for assistance, and ask a handful of trusted friends or mentors to read your résumé and offer suggestions.

Your community also offers resources at workforce centers and Veterans Administration Offices. Check with veterans groups on LinkedIn for help, and don't forget to check out your local library—they have a plethora of books and materials on résumé writing. Translating your military experience and skills into civilian language is a crucial step in drafting a first-rate résumé .

Finances/Savings
Some are fortunate to have employment secured before they retire from the military. Others have a gap between retirement and their first job in the civilian sector. Plan for this gap well ahead of time, especially if you are married; military spouses will often find themselves without work as they transition to a new location. Try to have three to four months (or more) of savings to cover financial obligations during this gap. Come up with a budget and stick to it—know your financial requirements on a monthly basis.

Because I retired as a drill-status Guardsman—in other words, I didn't have 20 years of active service—I won't receive a retirement check until I turn 60. Fortunately, Gary and I had prepared to live on one income. We had no debt and had saved for this major transition. Not taking credit card debt into retirement gives you freedom and flexibility; conversely, having debt when retiring can cause extra stress on a marriage as well as on single individuals. If you get behind in payments, make sure you are working with your creditors. Save and prepare for your retirement—it creeps up faster than you think.

Housing

A year before my husband retired from the Marine Corps, we decided we would remain in Hawai'i. Not having to move or look for a house was a great relief. My transition, however, was a bit more complicated. About a year before my retirement, we had a conversation about where we were going to settle down and call home. We both had strong opinions about where we *didn't* want to live and decided that the Colorado Springs area met our criteria. We flew to Colorado for a weekend, met with a realtor, and did not find anything that fit what we both wanted in a home. At that point, we decided to build a house. I would not recommend building a house while you're living far from the actual build site. In hindsight, we should have relocated to Colorado and rented during the build to have eyes on the project.

Before you decide on a home—both the house and the city/state—consider these factors: proximity to family, taxes, job opportunities, quality of schools, and access to military health care, commissary, and exchanges. Decide if you will rent or purchase a home. If you are buying, don't forget your earned VA home loan benefits: you can purchase a home using your VA benefits with no money down, no private mortgage insurance, and low interest rates. Companies like Veterans United Home Loans can assist with this process.

Schools

If you have school-age children, thoroughly evaluate the quality of public and private education. Not all areas are the same—do the research and visit schools near your potential new home. Do an online search for best schools in your state, and consider factors that are important to you: what percentage of students go on to college, for example, or class size and average test scores. Look at school district websites, and involve your children in the search. Before you choose a private school, make sure your budget can accommodate the cost. Some private schools offer a payment plan, reduced tuition for military, or scholarships.

Wardrobe

More than a year before my husband's retirement, we began gathering pieces for his civilian wardrobe. Not knowing exactly what type of job he would get, we started with the basics. It's important to purchase classic pieces that are timeless. One month we bought several pair of shoes: black, brown, and cordovan. Later we added different styles, like lace-up and penny loafers. A month or so later, we bought trousers in classic styles and basic colors. Then came the shirts. In Hawai'i that was pretty easy, as most businessmen wear aloha shirts to work, but for the other 49 states he needed regular colored and solid dress shirts. The next month we purchased ties and belts to match. All men should have at least one or more suits for interviews—and do not wear military corfams as civilian dress shoes!

A wardrobe can be fairly costly, so it's a good idea to start early and purchase quality items slowly. You can find new or slightly used clothing at many affordable places like Clothes Mentor, Plato's Closet, Goodwill, or thrift shops on base. Many Army bases have an event called "Suits for Soldiers," where a Soldier can get a donated suit for free.

A wardrobe can become expensive if you shop at high-end stores—be smart and look for quality items at an outlet mall or a military exchange.

Having gone through this process with my husband, I knew what wardrobe I would need after I retired. I already had classic pieces of clothing: basic dresses, suits, blouses, shoes, and purses. As the need arose for something particular, I bought it, and as time went on, I added new articles of clothing and shoes. It's also important for both men and women to have a nice-looking dress coat for cooler weather.

An area that often gets overlooked in budgeting for a wardrobe is the cost of alterations. I recently took in an evening gown that needed to be shortened, and the alterations were almost as much as the gown.

Health
This was an area I neglected until six weeks before I retired, when I had two surgeries. Trust me, this is not the time to address your health. Make your well-being a priority and have medical and dental checkups months in advance of retiring. I thought I'd have one medical appointment and be on my way. One appointment turned into two and three . . . and it was months later before I was signed off. Make your medical appointments early and take care of your health.

Filing your VA Disability Claim
This is extremely important! Do not let this slip by and think it's not worth your time or energy. Gather all your medical records (civilian as well as military) and meet with a veteran service organization like the VFW or American Legion. You can find these volunteers through the Airman/Soldier Family Readiness Centers on base, or inquire at your TAP class. This is for your future care for injuries or health-related issues you sustained while on military service. You may even be eligible for some compensation. Whether Active, Guard, or Reserve, this is a must-do!

Vehicle Maintenance
Before you start on your road trip to your new home, make sure your vehicles are in good repair. Have a tune-up, check the tires, and have an emergency pack with the essentials.

*   *   *   *   *

As I planned for retirement, I decided I would not seek employment. I took a break for six months, settled into our new home, and looked through boxes that had not been opened for 25 years. After that, I went to college for a year to finish my undergraduate degree. I also visited the local Workforce Center and took a few classes that were free to veterans.

About a year after retirement, two national non-profit organizations asked me to serve as a trustee on their board of directors. United Through Reading (UTR) and Military Child Education Coalition (MCEC) are doing great work for military families in the areas of literacy, reducing separation anxiety, and building resiliency for military kids, as well as helping military families transition from school to school. I served one three-year term with MCEC and am on my second term with UTR. I'm currently serving on UTR's executive committee as the board secretary and on the strategic growth committee.

As a newcomer to Colorado, I decided to connect with the Colorado National Guard and attend their state conference. I served as the director of Colorado Air National Guard membership for two years.

Giving back has always been important to me. As scripture teaches us, "to whom much is given, much is expected." A friend invited me to an Employer Support of the Guard and Reserve (ESGR) social. As a Guardsman, I am well versed on the good work ESGR does for our military members and employers. I currently serve as a volunteer with the Colorado Springs chapter of ESGR.

In mid-2016, I joined the military relations team at Veterans United Home Loans (VUHL) as a military advisor. Under the tutelage of Pam Swan, we educate veterans across the country about their VA home loan benefits. Our goal is for every veteran to know and understand the valuable benefit they have earned through their service to this great country. We also educate the employees of VUHL to better serve the veteran community with their home-buying experience.

From early 2016 through mid-2017, I began writing a book chronicling my upbringing and military career. In August 2017, I published *From the Prairie to the Pentagon*. It was a long process with starts and stops, and in the end I am very proud of the book. It's the story of a young farm girl who attended a one-room country school, joined the Air Force, and finished her career having achieved the highest position ever held by an enlisted woman in the history of the U.S. military.

I continue to travel and speak at military and civilian events on leadership and the importance of professional and personal growth and mentorship. I'm proud to be part of an amazing group of leaders in a joint venture called Summit Six LLC. Ken, Mike, Rick, Jim, Skip, and I started this leadership consulting company in 2017 and are excited about the possibilities.

Gary and I live in Colorado and enjoy hiking, snowshoeing, and the great outdoors. In 2016, we even hiked our first "14er" (a mountain over 14,000 feet). We love our faith community at Our Lady of the Pines and attend various activities through the church. ∎

TOP: Country school, grades 1-5, in rural Little Falls, Minnesota, early 1970s.
ABOVE LEFT: First grade class picture.
ABOVE RIGHT: High school graduation, 1978.
RIGHT: Offutt AFB, Air Traffic Control tower training, 1985.

TOP LEFT: Airman Basic Jelinski, Basic Military Training, Lackland AFB, 1984.
TOP RIGHT: Cammied up in MILES Gear, California Air National Guard, 1987.
ABOVE: 154th Wing Commander Col. Peter Pawling pins new wing command chief, February 2004.
LEFT: On Saddam Hussein's throne in Iraq with Hawai'i TAG Maj. Gen. Robert Lee, 2009.

TOP: First sergeants of Hawai'i Air National Guard.
ABOVE: Memorial Day breakfast at the White House with heroes of the 442nd Infantry Regiment.
RIGHT: Hawai'i National Guard Birthday Ball with Gary and Ashley.

TOP LEFT: 26th Chief of the National Guard Bureau Gen. Craig McKinley.

TOP RIGHT: Delivering remarks before her retirement ceremony, June 2013.

ABOVE: Fighting 69th Infantry Regiment, New York Army National Guard, New York City, St. Patrick's Day 2013.

LEFT: Discussing the damage caused by Hurricane Sandy, New Jersey, October 2012.

Charles W. "Skip" Bowen

MASTER CHIEF PETTY OFFICER OF THE COAST GUARD

C harles W. "Skip" Bowen became the tenth Master Chief Petty Officer of the Coast Guard (MCPOCG) on June 14, 2006, and served until May 21, 2010.

Bowen spent the first eight years of his life in the small fishing town of Fortescue, New Jersey, before moving with his family to Florida. The family spent several years in Fort Pierce and then settled in Sebastian for Bowen's last year of high school. After attending basic training in 1978 at Coast Guard Recruit Training Center Cape May, New Jersey, his first duty station was to a patrol boat, the U.S. Coast Guard Cutter *Point Swift* in Clearwater, Florida. In 1980, Bowen was assigned to Coast Guard Station Marathon in the Florida Keys just in time for the Mariel boatlift, a six-month mass emigration of Cuban refugees to U.S. shores. A subsequent assignment at Station Fort Pierce, Florida, was followed by another patrol boat, the newly commissioned U.S. Coast Guard Cutter *Farallon*, out of Miami.

Bowen joined the U.S. Coast Guard Cutter *Point Arena* in the mid-Atlantic seaboard as the executive petty officer. Upon advancing to chief petty officer (CPO), Bowen was assigned as the officer in charge (OIC) of Station New Haven, Connecticut, in June 1990. He was transferred to Station Sand Key in Clearwater Beach, Florida, in 1994. In 1997, Bowen was assigned as the OIC of the U.S. Coast Guard Cutter *Point Turner* in Newport, Rhode Island, until her decommissioning in April 1998. He was then assigned as the OIC of U.S. Coast Guard Cutter *Hammerhead* based in Woods Hole, Massachusetts, the first of the high-tech 87-foot patrol boats on the East Coast.

From 1999 to 2001, Bowen served as the District Seven command master chief (CMC). In May 2002, he graduated with distinction from the U.S. Army Sergeants Major Academy. While at the Academy, he was selected as one of the few non-Army students ever to serve as a class vice president, and received the prestigious William G. Bainbridge Chair of Ethics Award.

From June 2002 to June 2004, Bowen served as CMC of the Headquarters units. In addition to those duties, he also served as the interim MCPOCG July–October 2002. In 2004, he was assigned as the OIC of Coast Guard Station Marathon in the Florida Keys, where he was responsible for search and rescue and enforcement of federal maritime laws and regulations for a third of the Florida Keys.

Bowen received a Bachelor of Science degree, magna cum laude, from Excelsior College in Albany, New York, and a master's degree in business administration (MBA), summa cum laude, from Touro University International in California.

His personal awards include the Coast Guard Distinguished Service Medal, four Meritorious Service Medals with "O" device, four Coast Guard Commendation Medals with "O" device, three Coast Guard Achievement Medals with "O" device, Commandant's Letter of Commendation, six Meritorious Team Commendation Awards with "O" device, two Coast Guard Unit Commendations with "O" device, seven Coast Guard Meritorious Unit Commendations with "O" device, and the Coast Guard Sea Service Ribbon with two stars.

Bowen's first position post-retirement was as a senior program manager with CACI, a company in Fairfax, Virginia, that provides information solutions in support of national security missions. During his time at CACI, he received his Project Manager Professional (PMP) certification. In April 2011, Bowen began working at Bollinger Shipyards in Lockport, Louisiana, as the project manager of the Fast-Response Cutter program. The following year, he was promoted to vice president for government relations.

Since his military retirement, Bowen has served on the board or advisory council of several organizations affiliated with the military, including the Armed Forces Retirement Home, First Command Financial Services, Broward Navy Days, and Mission Readiness. He has served as co-chair of the Commandant of the Coast Guard's National Retiree Council and is currently the president of the Association for Rescue at Sea. He also works with Veterans United Home Loans as a military advisor.

Bowen and his wife, Janet, have four children and live in Sebastian, Florida. ■

Introduction: Future Guardian

Anyone can hold the helm when the sea is calm.
~ Publilius Syrus

I was still in high school the first time I visited the Coast Guard recruiting office in Winter Park, Florida. Graduation was looming, and I knew a life on the sea was the future I wanted. Most of my family members were seamen going back to the mid-19th century; my dad had served in the Navy before becoming a commercial fisherman, and my younger brother was already set to follow in his footsteps. The sea was in our DNA. I too had once entertained the idea of fishing for a living, and probably would have gone that route had I possessed the skill set needed to survive as a fisherman. I can honestly admit I did not. Marine life was in very little danger from me. The fish seemed to be the only ones that thought it was a good idea for me to continue in the family business.

Despite my lack of talent as a fisherman, I loved the ocean and still hoped to find a way to incorporate it into my life. I found the answer in a 1973 issue of *National Geographic* that had dedicated several articles to the Coast Guard and its heroes. From that point on, I set my sights on joining the nation's smallest and most versatile military service—and the only one outside the Department of Defense.

At the recruiter's office, I learned there was a yearlong wait list to join the Coast Guard, and if I were truly interested in joining, I would have to make monthly calls to express my continued interest and keep my place on the list. I was faithful about making those calls for six months, doing my best to make a living as a fisherman in the meantime. Unfortunately, for some reason I missed the seven-month call. When the time for the eight-month call rolled around, I diligently dialed the recruiting office and spoke to my recruiter, FS2[1] Benny Gravitt, who informed me that because I hadn't called in the previous month, I had been dropped to the bottom of the list. I would have to begin my yearlong wait from scratch.

Much to the chagrin of the fish, who I'm pretty sure considered me some sort of defender of oceanic life, I had no interest in struggling through another whole year of fishing. I was ready to start my career, and I made that perfectly clear to the recruiter.

"Look," I told him, with all the indignation my 19-year-old self could muster. "I have options! I can go join the Navy, and they'll have me sworn in today!"

God was looking out for me, because instead of suggesting that I do just that, the recruiter put me on the delayed-entry program, and I was sworn in the next day.

In May 1978, I entered recruit training at Cape May, New Jersey, joining Recruit Company Uniform 100. I will never forget entering the classroom and seeing a foreboding message on the chalkboard: "You are standing into danger." I remember those words caused me some concern about what I was getting myself into. I later learned that each

---

[1] Food service specialist petty officer 2nd class

letter of the Coast Guard phonetic alphabet is associated with an international maritime signal flag, which in turn is associated with the recruit company bearing the same designation. The flag for the letter U ("uniform") signifies the warning, "You are standing into danger."

Despite any last-minute reservations, I persevered and discovered recruit training, or boot camp, was challenging but survivable. I learned what it meant to be a Coast Guardsman and how important it was to function not simply as an individual, but as part of a team wholly dependent on the cooperation of each member. Though grueling and demanding, the training and education we received at boot camp gave my fellow Coasties and me the skills to keep ourselves alive—more importantly, we learned how to work together "so others may live."[2]

I enjoyed my service in the Coast Guard and moved up steadily in my career, but I didn't entertain the thought of becoming the master chief petty officer of the Coast Guard (MCPOCG) until I was advanced to chief petty officer in 1990. I was attending the Chief Petty Officer (CPO) Academy when I became good friends with Senior Chief Vince Patton, who eventually became the eighth master chief petty officer of the Coast Guard. He spoke to me about his goals to become the MCPOCG one day and even showed me his class ring from his own days at the CPO Academy. He had altered the ring, changing the insignia of a chief petty officer to the three-star insignia of the master chief petty officer of the Coast Guard. His confidence and determination to reach the highest enlisted position in the Coast Guard made me realize the possibilities for my own career.

As a chief petty officer, I was assigned as the officer in charge at Coast Guard Station New Haven, in New Haven, Connecticut. My group commander (Regional Command) and captain of the Port Long Island Sound was Capt. Thad Allen. Captain Allen was an inspiring leader, capable of motivating and empowering everyone on his team. He was a visionary who effectively communicated and provided clear direction to execute his vision better than anyone I had ever served with. I was so impressed with him and confident in his exceptional abilities to lead that during my out-brief from Station New Haven in 1994, I told him, "Sir, when you become the commandant of the Coast Guard, I would like to be your master chief petty officer of the Coast Guard." Twelve years later, when Admiral Allen became the commandant of the Coast Guard, he selected me to become his MCPOCG.

Throughout my career, from seaman recruit to master chief petty officer of the Coast Guard, I've gleaned indispensable lessons in what constitutes good leadership and what qualities effective, admired leaders possess. I have studied the examples of the exceptional leaders I've had the privilege to serve with and have strived to incorporate those lessons into my own leadership style. Unfortunately, if you spend any length of time with an organization, you will run across individuals who have been placed in high-ranking positions despite their inability to provide strong, effective leadership. Some rule by micromanaging, belittling, and senseless shouting. My best advice when you find yourself answering to such a leader is to simply survive and outlive them. Let them be a lesson in what *not* to do, which you can apply to your own style when you step into a leadership role yourself. ■

---

[2] "So others may live" is the creed of the Coast Guard rescue swimmers.

Lesson One: Lead Boldly

The greatest leader is not necessarily the one who does the greatest things.
He is the one [who] gets the people to do the greatest things.
~ Ronald Reagan

G ood leadership requires boldness. Commit yourself to those you are leading, because the moment you take the command, you become responsible for them and their success. You are their mentor and advocate, and the exemplar of everything you expect from them. Your crew members rely on you to not only guide them but also to equip them to handle every situation.

The moment you step into that leadership position, your job is to make it clear to your crew members that they are part of a remarkable team. They are part of something bigger than themselves—something they can take pride in. As part of your crew, they become more than just co-workers. They are shipmates who will embark side by side on a multitude of missions where they will put their own lives in each other's hands. When one team member is in trouble, it is their teammates' duty and privilege to watch out for them, encourage them, and support them.

Your duty as their leader is to hold the team together and help everyone thrive, both as a unit and as individuals. Show them that as they fulfill your expectations of conduct and duty, you will remain committed to having their backs. Empower them to take ownership of their actions. Respect them. Not only will your crew members learn to respect you in return, but they'll also learn to respect themselves. This atmosphere of mutual respect cultivates trust between both parties, and trust is a catalyst for the success of your command.

The strength of the team is each individual member.
The strength of each member is the team.
~ Phil Jackson

I learned these vital lessons soon after graduating from recruit training, when I was assigned to my first duty station, Coast Guard Cutter *Point Swift* (WPB 82312). The *Swift* was an 82-foot cutter with a nine-man crew commanded by Master Chief Petty Officer Harry Turpin. I may have been fresh from boot camp with no real service experience yet, but I quickly learned that every crew member is vital to the mission—and that's especially true on the relatively small cutters and Coast Guard stations. The efficient, lean organization is one of the characteristics that makes the Coast Guard such a unique service; a high demand for professionalism and competency stretches to the lowest ranks.

My earliest experiences on the *Swift* taught me not only how crucial my training was but also how important it was for the crew to rely on each other and depend on the leadership of MCPO Turpin. The men and women of the U.S. Coast Guard are responsible for many missions, spanning from search and rescue to maritime law enforcement to national defense, and often these missions require them to enter life-threatening situations.

On one such occasion, the *Swift* was underway on a search-and-rescue mission near the Dry Tortugas, responding to a distress call from a sinking shrimp boat with several crew members on board. When the *Swift* arrived at the shrimper, the seas were manageable; a few of us easily crossed from the bow of the cutter to the stern of the shrimp boat, which was still taking on water. Our first priority was the safety of the shrimper crew, but after we had evacuated them to the cutter, we turned to our next task: saving the impaired vessel. While we struggled with dewatering pumps clogged with debris from the vessel's bilge, the weather conditions had been steadily deteriorating. Twelve-foot seas prompted Master Chief Turpin to make the call to abandon the sinking vessel and return to the *Swift*.

Our situation became something of a catch-22. The *Swift* was equipped with a small rescue craft known as a cutter boat, which had to be hoisted from the main deck by means of a single-point davit and then swung over the side by block and tackle and lines. Had we not been facing 12-foot seas, we would have launched the cutter boat and abandoned the shrimper. Unfortunately, the dangerous conditions that made it impossible to launch the cutter boat safely also made it nearly as impossible to navigate the *Swift* close enough to the shrimp boat to retrieve us. Both vessels were being tossed around in the waves, but Master Chief Turpin had no other choice but to guide the *Swift* as closely to the shrimper as he could while avoiding a collision.

With a level of seamanship that can only result from decades devoted to sea service, Master Chief Turpin positioned the bow of the cutter close to the stern of the shrimper, taking us off one at a time as the seas heaved both vessels up and down. I was the last one remaining on the sinking vessel, waiting for my chance when the gap between the cutter and the shrimp boat was short enough to cross. As soon as the bow of the 82 came within reach of the stern, I jumped. A fellow crewman grabbed my arm and hauled me aboard the cutter as the bow plunged down, narrowly missing the stern of the shrimper. Master Chief Turpin navigated safely away from the ill-fated boat, having successfully saved not only the lives of the shrimp boat crew but also the lives of his own men.

I have never encountered anyone who could handle an 82-foot patrol boat like Harry Turpin. His unmatched boat-handling skills made him a great boatswain's mate, but his devotion to duty, levelheadedness, and ability to unite his crew for a common goal made him a great master chief. The powerful sense of team unity I learned from him stayed with me for the remainder of my career, and in later years I tried to re-create that in every one of my commands.

> *Alone we can do so little; together we can do so much.*
> ~ Helen Keller

It's your job as a leader to not only unite your team but also to understand your crew members and find out what motivates them. There is a key for every group and individual that will unlock that motivation and allow them to excel; your job is to find the key.

When I became the officer in charge at Coast Guard Station Marathon in the Florida Keys, understanding my crew's specific challenges was key to my effective leadership. The small station had a crew of 40 Coast Guard enlisted personnel with pay grades ranging from E-2 to E-7. Despite its size and relatively junior personnel, the station was responsible for a large mission, including search and rescue, federal maritime law enforcement, and security of the U.S. maritime border for about a third of the Florida Keys.

Several years before I arrived, Station Marathon had struggled with numerous problems. As a result, the entire command was relieved for cause, and a new officer in charge (my predecessor) was brought in to basically clean up the mess. To do that, he felt he needed to use a heavy hand to restore and maintain order at a station that had been through so much disorganization. He was a good man, and since he was brought in to clean up a difficult situation, I'm not going to criticize his methods. However, I chose to do things differently. I wanted to communicate to the crew a new vision of how to move forward and leave the problems of the past exactly where they belonged—in the past. An opportunity to do just that arose during my relief week prior to taking command.

When I arrived at the station, I faced two challenges: a less-than-optimal command climate and a local culture that posed many challenges to young, mostly single crew members. After looking at the punishment log, I jokingly asked the executive petty officer for a list of people who *weren't* on restriction, since that list would have been far shorter. During our relief week, one of the young, non-rated personnel ("Fireman Smith") was caught doing 90 mph in a 35 mph zone at night on Big Pine Key. He was arrested and thrown in the county jail overnight.

Reluctant to leave me with a mess at the start of my command, my predecessor wanted to take the fireman to a Captain's Mast, a type of proceeding used to determine nonjudicial punishment. Besides restriction, extra duty, various monetary implications, and reductions in rate, nonjudicial punishment can delay attendance to "A" School, for which Fireman Smith had been in line prior to his offense. Graduation from "A" School would mean a promotion for the young man, so the implications of being prevented from attending were significant.

Instead of seeing the situation as a mess, however, I saw it as an opportunity to communicate to the crew that things would be operating differently under my watch. The Wednesday after my change of command ceremony, we held a Captain's Mast. The room was set up in the traditional manner, with a long table at the front covered with a green tablecloth, and a copy of the *Uniform Code of Military Justice* and the *Manual for Courts-Martial* sitting on top. Although the proceedings were nonjudicial in nature, we could call witnesses and take statements. Often, to make a point, commanding officers would have the whole crew attend the proceedings. For Fireman Smith's Captain's Mast, I did just that: I instructed the entire crew of Coast Guard Station Marathon to attend.

That morning, Fireman Smith arrived at the station with a large suitcase packed with his clothes and necessities. He knew the precedent, and he fully expected the Captain's Mast to result in being restricted to the station and prohibited from returning home for at least two weeks.

At the appointed time, we brought in the accused and instructed him to stand before the table. I paused, making a point to look at both manuals. I then looked up and ordered Fireman Smith to turn around and face the assembled crew. I chose my next words carefully. I wanted to convey the gravity of having constant and ongoing punishments, which is clearly what had been happening at the station for some time. The negative atmosphere of penalties and reprimands wasn't just bad for the individual being punished—

it was ultimately harmful to the mission, because it contributed to poor morale and disrupted unit cohesion.

I had learned that Fireman Smith was an excellent performer, so I asked the crew why he was now standing in front of me at a Captain's Mast during my first week. I explained that after reading the unit punishment log, I noted that nearly a third of the crew was either currently on restriction or had recently been on restriction for various infractions. I revealed that upon reading the investigator's report, I had discovered that two petty officers were in the vehicle with Fireman Smith. No alcohol was involved, just inexcusably poor judgment.

Big Pine Key, the scene of the infraction, is home to a population of endangered miniature deer called Key deer, the protection of which the locals take seriously. Driving at over 90 miles an hour is dangerous and reckless anytime, but doing it at night along a road known for frequent Key deer sightings was even more foolhardy.

But those weren't the points I wanted to make to the crew. What I wanted to highlight was that nobody else in that vehicle, including the two petty officers, had told Fireman Smith to slow down. Nobody pointed out that he was endangering the lives of other motorists on the road, those in the vehicle with him, and even the Key deer. The personnel of Coast Guard Station Marathon made up a team, but there was no evidence of teamwork in that vehicle. Nobody was watching out for anybody else, and nobody was concerned that Fireman Smith was making a decision that would have a huge impact on his future career.

After I heard the testimony of the witnesses, I instructed both petty officers in the car that night to join Fireman Smith, and I asked all three to face me. I told the petty officers that I was even more disappointed in them than I was in Fireman Smith. They were senior to him, and it was their responsibility to mentor and inspire him. I turned to Fireman Smith and sentenced him to several days of restriction, and then I immediately suspended the sentence. I wanted to drive home the point that things were changing. I announced to the crew that I expected more from them. We were going to be professional at all times, on and off duty. We were going to start watching each other's backs. We were going to start making sure that each of our fellow shipmates stayed out of trouble.

The crew of Coast Guard Station Marathon left the room that day with a new sense of purpose, identity, and unity as a team. They also had a promise that I would be looking out for them, just as they looked out for each other. I remained at that station for two years, and I never again had to hold a Captain's Mast. ■

Lesson Two: Lead by Example

Always do everything you ask of those you command.
~ General George S. Patton

L eaders should have the courage to practice what they preach. You can't ask those you supervise to do things you wouldn't—or haven't done in the past when you were in their positions—and expect them to happily, blindly oblige. When you take on a leadership role, you immediately step into a spotlight that illuminates and magnifies your actions, attitude, and experience. The way you choose to lead determines the way your crew will follow. Your crew is your mirror, and their behavior and performance will be a reflection of yours. Leadership is substantially more than issuing orders and reciting empty rhetoric to those serving under you. Leadership requires that you *lead*. You are the one blazing the trail, responsible for making decisions that influence the safety and effectiveness of your entire team.

When I was the command master chief (CMC) of District Seven,[3] I visited two units one day. In the first and smaller unit, the chief petty officer believed that his government housing allowance didn't come anywhere close to the amount that would afford him decent quarters in the area. The unfairness of his housing allowance dominated our conversation throughout my visit. Interestingly, his entire crew also expressed dissatisfaction with their housing allowance.

Later that day I visited another, larger unit located within a few miles of the first. All personnel from both units had the same housing allowance and lived in the same areas. As I visited the second unit, I asked the crew what they thought of the allowance rates. Not one person had a problem; the general consensus was that the rates were fair. Interestingly, the chief of the second unit also thought the allowance was fair. It was evident that the difference in the two crews' perception of the housing allowance rates could be tied directly to the attitude of their chief petty officer. Not only will your crew notice the tone you set and the attitude you convey, but these will become the standard by which they behave.

Transparency of information breeds self-correcting behavior.
~ Adm. Thad Allen, Commandant of the U.S. Coast Guard

When I was a young petty officer, I served under a chief who was a weak leader. The unit was characterized by poor performance and low morale. Crew members were constantly looking over their shoulders for fear of being admonished for any and every minor problem. The senior petty officers fell easily into the chief's leadership mold, using the precedent he set as a template for their own leadership styles. They were constantly shouting and belittling everyone's efforts, while offering little guidance or providing feedback on how to improve.

---

[3] Regional command responsible for the Southeastern United States, including Florida, Georgia, and South Carolina

Nor did they ever properly communicate what we were doing wrong—it seemed they just enjoyed the power trip they got by constantly berating us.

Instead of running a consistent, solid unit, the petty officers allowed for its steady decline. We received a pretty good evaluation on a district inspection once, but it wasn't due to the petty officers' successful management of the unit. Instead, it was because on the night before the inspection, the entire crew worked literally all night, cleaning and re-cleaning for a disorganized and demoralizing leadership team. I distinctly remember having to clean the same painted pipe in a storeroom over and over again as different supervisors kept coming through to inspect my work. It was four o'clock in the morning when we finally finished.

Another time at that same unit, one of our coxswains had a mishap with an old 40-foot steel hull Coast Guard vessel. By the time he brought it into the station, it was in a sinking condition. The exhausts were well underwater, with only a few inches of freeboard at the stern. We moved quickly and had several dewatering pumps going within minutes. We were feeling good about getting things under control and saving the boat, until the chief showed up. He stood on the dock screaming at us that we had better save his boat. I remember thinking, *If you're not going to help, I wish you'd just shut up and go back home.* Unsurprisingly, not many re-enlisted at that unit; leaving was cause for celebration. The senior petty officers followed the chief's lead and had the same negative attitude toward the crew, causing low morale and cultivating a sense of bitterness in all of us who were forced to endure their less-than-inspiring leadership tactics.

Your attitude, not your aptitude, will determine your altitude.
~ Zig Ziglar

On the other end of the spectrum, one man early on in my career made a lasting and significant positive impression on my life and on my leadership style. Chief Warrant Officer John Webb was the commanding officer at Station Fort Pierce when I was transferred there in 1983 as a young BM2,[4] and he taught me what true leadership really is. Without micromanaging, CWO Webb was involved in everything that occurred at the unit. He was one of the most proactive, positive leaders I have ever met. He was a teacher, counselor, mentor, disciplinarian, and part-time parent to a bunch of us part-time kids. He believed that every person and every situation was different, so "stock" answers found in regulations sometimes didn't apply. He taught us that supervisors should use their experience and judgment when dealing with individuals, that we should weigh issues and find a way to say "yes" if the situation dictated it.

CWO Webb valued the human element found in every situation and believed in accounting for it, because if a leader ignores that human element, morale, safety, and operations can suffer. He also knew how to communicate—he didn't believe in barking orders and screaming at station personnel for every little matter. In fact, if he ever raised his voice, we knew something major was happening and we had better listen closely and be ready to leap into action.

---

[4] Boatswain's mate 2nd class petty officer

I could write an entire book on John Webb and his methods of leading, mentoring, and managing, but one story fully captures his boldness as a leader and shows why we were all willing to follow his example. I was serving as the officer of the day one evening when we received word that several smugglers might be offloading multiple tons of marijuana near Wabasso Beach, about 20 miles north of Fort Pierce. The primary crew got underway in a 41-foot Coast Guard vessel. CWO Webb came into the station and took out an additional crew on a 21-foot Boston Whaler—much smaller and not meant for far-offshore use.

The Coast Guard vessels coordinated their approach to Wabasso Beach with other law enforcement organizations so that when they approached the beach from the water, the land-based units approached the offload site by road. Multiple vessels offloaded bales of marijuana over the beach to waiting vehicles. One of the vessels was a small fast boat, and when law enforcement approached, the vessel powered up and fled the beach straight offshore. Without hesitation, CWO Webb followed in hot pursuit.

In those days, Coast Guard vessels such as that whaler were outfitted with only a marine radio for communication and a compass for navigation. The vessel they were chasing was fast, but CWO Webb was able to keep them in sight. However, the further they went, the poorer communications became, and to make matters worse, the Whaler did not hold the amount of fuel required for a long, drawn-out chase. When they went past 20 miles offshore, I started to become quite nervous. We had asked for air support, and a Coast Guard helicopter was on the way, but it was still some distance off. When the chase hit 25 miles offshore, our regional command in Miami called and asked how it was going and exactly where our boat was. I had to tell them the Whaler was about 25 miles directly off Wabasso Beach, but it was OK because "our commanding officer is aboard."

I don't think anyone felt calm about the situation, but seas were smooth and the helicopter was near. As the two vessels approached 35 miles offshore, the helicopter arrived and angled in front of the fleeing drug smuggler, allowing CWO Webb and the crew to board the vessel and arrest the smugglers. Ironically, the Boston Whaler had to be towed back by the seized smuggling boat when it ran out of fuel. Had the smugglers been aware of just how close to escaping they really were, I'm sure they would have held out a little longer. Only John Webb could have pulled that off.

My time serving with CWO Webb gave me invaluable experience and molded me into the leader I later became. Semper Paratus ("always prepared") was our motto, and we practiced "total quality management" long before it became a popular business-school management approach. Everyone watched everyone else's back. We were a team. John Webb's unit was characterized by exemplary performance and high morale, and any one of us would have followed him to hell and back. Not surprisingly, almost everyone at that unit re-enlisted. The petty officers followed CWO Webb's example and were motivated, respected leaders. We all felt that each member of the crew had something important to contribute to the team, which led to a great deal of job satisfaction.

In both examples, leadership characterized these units. The first was subjected to toxic leadership that poisoned every aspect of the job. Morale was low and unit cohesion nearly nonexistent. The second unit thrived under the leadership of John Webb, and all who served under him worked hard to meet and even surpass his expectations. He led by example and cultivated an atmosphere of unity for each of us on his team. ∎

Lesson Three: Situational Awareness

Before you are a leader, success is about growing yourself.
When you become a leader, success is about growing others.
~ Jack Welch

There is no limit to what a man can do or where he can go
if he does not mind who gets the credit.
Plaque on President Ronald Reagan's desk in the Oval Office

Through innovative thinking, situational awareness, and appropriate risk management, good leaders know how to make things happen. One-dimensional, narrow-minded thinking can be the downfall of a leader—and if they fail, so do their crew and the mission. It's easy to play it safe during an operation by closely following regulations and thinking well inside the box, but the leaders who take the initiative and recognize that we don't live in a one-size-fits-all world are the ones who get things done. Find the best solution, and as long as it's not illegal, immoral, unethical, or clearly against policy or regulations, implement it.

From 1986 to 1989, I served as the senior boatswain's mate on board the Coast Guard Cutter *Farallon* (WPB 1301), stationed in Miami Beach. She was the first of the 110-foot Coast Guard patrol boats built by Bollinger Shipyards and bought by the Navy as part of then-Vice President George H. W. Bush's initiative to build up Coast Guard assets in the Caribbean drug war.

In 1988, the *Farallon* was involved in an operation to bring three large drug-smuggling vessels back to the United States for prosecution. Towing even a single vessel with one of the first-generation 110s was difficult; towing three was exceptionally challenging. These cutters were equipped with large Paxman diesel engines—the same used in trains in England—and had tremendous power. When a Class A 110 clutched into gear, even with only one of its shafts, it attained a nine-knot speed forward within seconds. This quick acceleration often posed problems when towing vessels, particularly in rough weather. In fact, so many drug vessels had been sunk because of damage inflicted during the towing process that the *Miami Herald* wrote an exposé accusing the Coast Guard of sinking the vessels on purpose.

Nevertheless, the *Farallon* was tasked with towing three large trawler-type vessels back to Miami from the southeast Bahamas. The safest, easiest option would have been to just tow one of the vessels and ask other WPBs to tow the remaining two. However, if we saddled two additional Coast Guard boats with our responsibility, they wouldn't be available for other critical missions.

After careful consideration of the pros and cons of each option, the leadership team decided to undertake a tandem tow of all three vessels, even though one of them was already taking on water. We knew it would be a difficult, complicated journey, but doing it solo was still more favorable than the alternative. Thanks to some innovative problem solving, we successfully towed three mammoth drug-smuggling vessels for roughly 500 miles from the southeast Bahamas to Miami. Had we simply relied on the standard operating procedure, the task would have been impossible.

We had to be willing to consider unorthodox solutions. Towing just one vessel required a towline attached to the stern of the 110 and an enormous bridle, which connected the tow line to the vessel being towed. Most of our bridles were made of steel-wire rope and formed the shape of a Y; attached to the towline on one end, it split into two legs on the other. Each leg was attached to the bow of the disabled vessel, one on the port side and one on the starboard side. Our crew only carried two large bridles, so we had to fashion a third bridle out of three-inch, double-braided nylon line. (Steel-wire rope bridles are recommended for large, trawler-type tows because any other type of line tends to chafe through.) Using every one of the *Farallon's* large towing hawsers, we were able to get underway with all three vessels, preventing the third, impaired trawler from sinking by pumping it out every few hours. With approximately 400 to 700 feet between vessels, we presented quite a spectacle as we made our way through the Bahamas.

Large amounts of free-floating marijuana in the bilges of the third, waterlogged vessel kept clogging up the pumps every time we tried to de-water the vessel. Among the cargo of contraband drugs and other oddities, there happened to be a live rooster in a chicken-wire cage. One of our crew came up with the idea to "MacGyver" a secondary strainer around the pump suction from the chicken wire. The plan worked like a charm, and from that point forward we were able to keep the level of water in the bilges at an acceptable level.

As the senior boatswain's mate on board, I was tasked with any towing issues, as well as some of the more difficult ship-handling maneuvers. During one of the last nights underway, we were slowly moving through the New Providence Channel when conditions began to grow rough. We were facing six- to eight-foot seas when I was called to the bridge with an emergency: the nylon towing bridles had chafed through and snapped, and the last vessel in tow had broken free.

I immediately needed to do two things. First, we had to jury-rig a new towing bridle on board the third vessel, and second, we had to negotiate our second towed vessel close enough to the bow of the third vessel so we could get a tow line back to it. After we discussed the evolution thoroughly, I took a couple of crewmen in the small boat to transit to the adrift third vessel and create a makeshift bridle on the bow.

Once on board the vessel, we crafted a new bridle and did our best to compensate for the weaker, nylon line by wrapping plenty of chafing material around all points on the line that would rub when we got back under tow. Then we stood by as the *Farallon*, still towing the other two drug-loaded vessels, repeatedly tried to maneuver the second vessel in tow close enough to get a line to us. High seas and limited visibility increased the difficulty of the already complicated operation. We hooked up the tow after several failed attempts, and with the new bridle in place, we completed the passage back to Miami safely.

Acting alone, even the smartest, most experienced leaders rarely find the most optimal solution. Inspire your team to come up with the best solution and empower them to think creatively. In the case of our three-vessel tow, the commanding officer of the *Farallon*, then-Lt. Missy Wall, enabled all of us to contribute to what was ultimately the best resolution and got the job done.

The commander in the field is always right and the rear echelon is wrong, unless proved otherwise.
~ Gen. Colin Powell

In 1999, shortly after I became the command master chief for District Seven, six Cuban migrants tried to row to Miami Beach in a wooden skiff. The national media dubbed the incident "Surfside Six." Coast Guard Station Miami Beach dispatched response boats to intercept the Cubans and determine if they were indeed migrants illegally trying to enter the country. Once they had that confirmation, the Coast Guard's task was to apprehend the migrants prior to reaching land. This was in accordance with the so-called "wet foot, dry foot" U.S. government policy: if Cuban migrants were caught at sea, they were to be detained and repatriated to Cuba; however, if they reached dry land, they would be allowed to remain in the United States as immigrants. Adding furor to an already difficult operation, several news helicopters were already hovering in the air just up the beach, covering a press conference held by Vice President Al Gore. They caught wind of the migrant situation almost immediately, and news stations began broadcasting live coverage of the entire event.

Coast Guard crews only have a few options when deciding how to detain a small, wooden rowboat filled with migrants. With the eyes of the entire country watching them on live TV, the Miami Beach crew began attempting to bring the migrants into custody. They first tried pepper spray, hoping to temporarily incapacitate the migrants with minimal force, prevent their continued movement toward the beach, and bring them on the Coast Guard vessel where they would be detained and sent back to Cuba. Unfortunately, the pepper spray merely aggravated the migrants and inflamed the situation even more than it already was. Next, the Coast Guard crew tried to get a line on the skiff so they could tow it back out to sea. That failed, so they attempted to use fire hoses to push the boat away from the shore.

At that point, the migrants jumped into the water and swam wildly for the beach. In a desperate move, one of the men threatened to injure himself by swimming into the propeller of the Coast Guard boat, prompting one of the crew to once again use pepper spray to stop him.

While the chaos off Miami Beach was escalating, so were the angry protestations of Miami's large Cuban-American population on land. Thousands of protestors gathered for demonstrations outside the Coast Guard station, eventually forcing authorities to shut down MacArthur Causeway, a busy six-lane thoroughfare, during rush hour.

As tensions continued to mount, BM1[5] Bradley Klimek from Station Miami Beach arrived at the scene. As the senior officer on scene, he quickly evaluated the swiftly deteriorating situation, then called his operational command in Miami to recommend immediately ceasing the use of pepper spray against the swimmers. The response he received from command, however, was "continue to use minimum force necessary to compel compliance." The directive was straight out of Coast Guard policy handbook, as every action of the crew had been up to that point.

---

[5] Boatswain's mate 1st class petty officer

As instructed, the crew continued employing the same tactics without success. The handbook policy, in this case, was causing a litany of problems not only for the Coast Guard crew and migrants on the scene, but also for the entire population of Miami. The huge crowd of angry protestors was growing quickly, and a fire started when some of them became violent. MacArthur Causeway was still shut down, and the protestors were spilling over to nearby highways.

Frustrated, BM1 Klimek called his operational command again. This time his request went all the way up to the District Seven Command Center, which simply echoed the previous response: use minimum force necessary to compel compliance. After receiving the by-the-book answer for the second time, Klimek took matters into his own hands, making an on-scene command decision to stop using the pepper spray. Two of the migrants reached the beach and were taken into custody by U.S. Border Patrol, while the remaining four were pulled out of the water by the Coast Guard. Later, in the wake of massive protests and outcry, all six migrants were released and afforded legal residency in Florida.

A full investigation reviewed the operation and the Coast Guard's use of force that day. The crew members were cleared of all wrongdoing; investigators determined they had been following prescribed policies and direct orders. Coast Guard policy, however, changed after that incident; crews could no longer use pepper spray against individuals in the water unless they were presenting a direct threat to the safety of the Coast Guard or themselves.

The Surfside Six events taught me two important lessons. The first is to recognize when policy is not working and be willing to take alternative action. The second is that there will be times when a leader is far removed from the action at the nucleus of the operation. Your interpretation of the situation is clouded by distance. The success of your mission and the safety of your people depend on your ability to listen to those in the field who may have a much clearer understanding of the situation. Trust the crew members on site to reset and respond accordingly.

In the case of the Surfside Six, the poor decisions at the top not only affected the immediate mission but also led to significant fallout for Station Miami Beach, District Seven, and the entire Coast Guard. For the next several months, I spent many hours responding to questions from concerned Coast Guard personnel throughout the service. Some believed the actions of the crew were unacceptable, while others were concerned that leadership would use the crew as scapegoats for the public backlash.

I also spent a great deal of time at Station Miami Beach dealing with the aftermath of the incident. Although the crew members had done everything right according to policy, they were concerned they were going to get thrown under the bus as a response to all the public outrage. Even the Department of Justice had launched an investigation. This could all have been avoided had someone in leadership had the presence of mind to stop quoting policy and give clear direction earlier in the case. Ultimately, BM1 Klimek was honored for his actions that day and awarded the Navy League's prestigious Douglas A. Munro Award for outstanding leadership and professional competence.

Only a few months after the Surfside Six events, we were involved in another high-profile Cuban migrant case. Fishermen found a five-year-old boy named Elián González floating unconscious in an inner tube, 60 miles off the coast of Miami. Elián's mother and 10 other Cubans had drowned when their small aluminum boat capsized in a storm on their way to reach U.S. soil. Elián was the only survivor. His mother had placed him in a rubber tube to keep him afloat, and he was dangerously dehydrated and suffering from hypothermia when the fishermen found him. Per standard operating procedure, they turned the boy over to U.S. Coast Guard custody.

Adm. Thad Allen, the commander of District Seven, had two choices. He could blindly follow policy and hold a severely dehydrated child offshore until a decision could be made whether or not to send him back to Cuba, or he could make a judgment call to set policy aside and save a little boy's life. Without hesitation, Admiral Allen set policy aside and ordered the Coast Guard to bring Elián to U.S. soil to receive lifesaving treatment.

Although policy dictates the appropriate course of action for most situations, good leaders should not be afraid to set policy aside if the situation dictates. Typically there are waiver processes that allow a leader to modify policy to fit a given situation—learn them. A good leader knows how to identify a unique set of conditions under which blindly following policy would cause more harm than good. Examine the situation from the perspective of your team members on the front lines. No matter how much experience you have, rely on your team to make sound judgment calls. Once you have a clear understanding of a situation that doesn't fit "by the book" standards, you have three immediate jobs. First, make a decision to depart from standard operating procedures. Second, determine an acceptable, preferable alternative to stated policy. Third, seek and implement a waiver to achieve that alternative. ■

Lesson Four: Leadership and Risk Management

Take calculated risks. That is quite different from being rash.
~ Gen. George S. Patton

There will be times in your career when you must make quick decisions in the heat of the action. Risk management can be an essential tool in these situations, but find a balance between taking absolutely no risks and taking unnecessary risks. Too often, leaders can become so risk averse that they actually do nothing, which can result in mission failure or even the death of a crew member.

During my time in the Coast Guard, we used a popular risk-management model known as GAR (green, amber, red). Leadership evaluated six factors of an operation for risk, including on-scene weather, operation complexity, and fatigue. Each factor had a point value between zero and ten: zero for no risk, and ten for maximum risk. The final total of all the point values indicated the overall risk of the mission. Zero to 23 indicated green, or low risk; 24–44 indicated amber, or moderate risk; and 45–60 indicated red, or high risk. If an operation was classified red, a negative outcome was likely.

No operation in the maritime environment is ever entirely risk free, but leadership should consider the level of risk in each case and then attempt to bring that level down. Of course, sometimes operating parameters can't be budged and a risk level remains high. I have known some leaders who use a high-risk level as an excuse to walk away from the evolution and do nothing. If you've evaluated all factors and you simply can't lower the risk level, there are times when you should still continue the operation. Risk-management principles should inform action, not dictate it.

Once you've made the decision to operate in the red, it is all the more crucial to remain within standards. While you can depart from procedure (with a waiver) for a moderate- to low-risk operation, the same is not true for high-risk operations. Departing from standards in such a perilous environment will almost always result in disaster. As I learned in one particularly risky case, never let the heat of the moment affect your ability as a leader to remember established protocol and make solid, thoughtful decisions.

The events occurred in 2004, while I was the officer in charge at Station Marathon in Southern Florida. During that time, between 2004 and 2006, a large influx of Cuban migrants attempted to make it to Florida in numbers just short of a mass migration. Many of them took their chances on rafts—slow chugs that were barely seaworthy—and sometimes even on cars and trucks that had been adapted to float. If a migrant had the money, however, they would often pay smugglers to transport them the 90 miles from Cuba to the United States on boats we called "go-fasts." Loaded with 20 to 40 migrants, these vessels were typically 30 to 45 feet in length and powered by multiple large outboard engines. The ill-advised "wet foot, dry foot" policy led to many dangerous situations for the Coast Guard, as Cuban migrants often became aggressive when they sensed dry land was near. I firmly believe this policy led to thousands of deaths over the years, as desperate Cubans took to the sea in search of freedom in unseaworthy craft and ultimately drowned in the Straits of Florida.

One evening in late 2004, Station Marathon got a call from Station Key West, the regional command for the entire Florida Keys. One of their crews had been pursuing a migrant smuggler operating a go-fast, and the boat had altered course toward the middle Keys, our area of responsibility. We promptly got our own crew underway in a 25-foot response boat. As the chase continued, the smuggling vessel refused the order to stop and intentionally rammed the Station Key West vessel, partially disabling one of the outboards. Our crew from Station Marathon immediately assumed the primary pursuit. Once the go-fast had shown hostility by ramming the Coast Guard vessel, the leadership at Sector Key West focused on stopping the go-fast using any means necessary. After a quick risk assessment, leadership determined to complete this operation in the red.

Joel Laufenberg, then a BM2 and at the helm of Station Marathon's vessel, vividly remembered the unique circumstances.

"Arriving on scene, I came along the starboard side of the vessel and noted three males on board," he recalled in a personal email. "There was one at the helm, one at the bow, and one on the starboard quarter. I believe it was about 28' long and a twin outboard vessel. The odd thing about this situation was that almost every other noncompliant vessel we had ever encountered had more than three people on board."

BM2 Laufenberg was pacing the go-fast when one of the smugglers on board threw a line at the Coast Guard vessel in an attempt to entangle the propellers. Laufenberg evaded the line. The smuggler retrieved it and tried again several times, but Laufenberg skillfully dodged the line each time.

"As I was pacing the go-fast, I was staying within 30 feet of his vessel, cruising at 30+ knots," Laufenberg recounted. "The wind wasn't blowing too hard, and the seas were flat, with the exception of a slow-moving, 1–2' swell.

"The go-fast was trimmed all the way up, so as it would climb a swell, the whole boat, with the exception of the engines, would come out of the water. As the boat re-entered the water, it was throwing a huge wave of water at us, soaking everyone outside of the cabin. Another tactic the operator was using was he would make sudden turns into me. Each time I would turn safely away, but because the turns he made were so abrupt, they would cause me to lose a bit of speed, and I would have to play catch-up for a minute or two to regain position. The operator turned at me like this multiple times throughout the chase.

"At this point there was quite a bit of anger slowly building amongst my crew. These guys were playing a serious game, and the chance of someone getting hurt was increasing with every moment. While I was regaining position off his starboard side after one of his attempts to turn into me, the crewman in the port quarter of the go-fast threw his line at us again. This time it came through the window of our vessel. I watched as MK3[6] Haeflinger struggled to get the line back out of the window, which was entangled dangerously around his hands. I did everything I could to ensure I stayed with the go-fast while MK3 freed himself and got the line clear of our boat and crew."

From the time the Marathon crew began chasing the smuggling vessel, they had also been requesting an SNO (statement of no objection) from Sector Key West for warning

---

[6] Machinery technician petty officer 3rd class

shots and disabling fire (WS/DF). Each crew carried two gun cases on operations: one held an M-16 rifle, which could be used with tracer rounds for both warning shots and disabling fire, and one held a shotgun, which was only authorized to be used for disabling fire. Sector Key West granted the request for WS/DF, and BM3 Joshua Soto quickly reached for the first gun case. He opened it and found a standard Remington 870 shotgun. He opened the second case, but it too contained a shotgun, much to the dismay of the crew. In the rush to get underway, the crew had grabbed two shotguns instead of a shotgun and an M-16.

BM2 Laufenberg, making a determined effort to exhaust all options, contacted Sector Key West to request permission to use the shotgun in lieu of the M-16 for warning shots. While waiting for Key West's response, he contacted the partially disabled Key West vessel and asked if they had an M-16 on board. They did, and Laufenberg's first instinct was to go get it. Before he could, Sector Key West granted the Marathon crew permission to depart from standard protocol and use the shotgun for warning shots—only the second time the Coast Guard had authorized use of WS/DF to stop a noncompliant vessel. The first such case had occurred only months earlier.

"It's hard for me to paint a clear picture of the chaos that was ensuing onboard," Laufenberg reported. "The go-fast wasn't letting me just come in and provide a stable platform for my gunner. He kept turning towards me, like he didn't want us to stop him. Weird. In training, the boat maintains a course and speed and lets us get our shots off—not in a real-life chase that night."

The smugglers continued their belligerent behavior, swerving the go-fast towards the Marathon crew and forcing BM2 Laufenberg to continuously engage in evasive action to avoid collision while BM3 Soto attempted to get his shots off. To make matters worse, the crew was repeatedly drenched every time the go-fast went over a swell; the gunner was baptized relentlessly with walls of seawater that saturated the shotgun. Laufenberg once looked back to check on Soto's progress, only to watch him dump large amounts of water out of the Remington's barrel. Laufenberg was turning so sharply to avoid the go-fast that he was often concerned he had tossed the gunner overboard.

Without the benefit of a harness to offer stability, another crew member, FN[7] Justin Malave, was doing his best to brace the gunner by pressing his knee into Soto's back, trying to forcibly hold him steady while he unloaded five rounds of copper sabot slugs into the water in front of the go-fast.

The smugglers were now unavoidably aware that the Coast Guard crew was armed, but the operator of the go-fast had no intention to end the chase. Once his continued noncompliance was evident, the Coast Guard crew began implementing the next step of protocol: disabling fire. BM3 Soto—still held as steady as possible by FN Malave—fired off five rounds at the starboard engine. Nothing happened. Laufenberg remembered that the report from the shotgun hadn't sounded quite right, but in the midst of all of the mayhem, he didn't have time to give it much thought. Five rounds of copper sabot slugs had been fired for warning shots, and another five rounds for disabling fire—this depleted all 10 rounds of available disabling fire ammunition.

---

[7] Fireman

The Marathon crew made the report to Sector, while tensions on board continued to escalate.

"We wanted nothing more than to stop this guy, and we couldn't," reported Laufenberg. "One small error, bringing the wrong armament, meant that we couldn't stop this guy. A radio transmission came to us from Sector, and we were told to 'shoot everything you got at the engines to stop the smuggler.' I recognized the voice on the other end of the radio, and I knew it was the voice of the highest-ranking officer at the Sector. Looking back, we should have questioned this order. But we didn't. We were too angry with the fleeing operator, with the situation, and with the way everything was playing out. We just received authorization to shoot whatever we had, so that's what we did."

What they had was 00 buckshot[8] loaded up into the shotgun. Laufenberg remembers two rounds fired into the engine. Once again, the action did not damage the engine enough to disable it. At this point, the crew received authorization to use LTL (less than lethal) munitions on the go-fast operator.

"Now it was really getting intense," Laufenberg said. "During the whole scenario, I was stationed off the starboard side of the go-fast. To this day I don't know why I did it, but I cut across the stern of the vessel and took station off his port side, as the BM3 got ready to take another shot. As I closed in on the go-fast, he began to make a wide turn to starboard. I turned with him to starboard. Then suddenly, the guy cranked the wheel hard to port, and I realized too late that I had just screwed up bad. There was no time for me to react. I had two crew members on the aft deck, and if I turned to port, they would be exposed and get hit. I did the only thing I could do—I came off the throttle, and the go-fast struck us just forward of the cabin on the starboard side. We were hit with pretty significant force and took major damage to the starboard side of the boat. I spun to port as he delivered the blow. I immediately tried to bring the throttles back up, but one of the engines wouldn't engage. He didn't hit anywhere near the engines, but he hit us so hard that the throttle linkage was dislodged, and we were out of the chase. Fortunately, by this time there were numerous other law enforcement assets on scene, and they followed the go-fast to shore and apprehended the three persons on board."

Not only were the three smugglers from the deck of the ship captured, so were the seven smuggled migrants who had been hiding out in the center console of the go-fast.

Laufenberg later discovered that the strange report he had noticed while Soto was firing disabling shots were from squib rounds. They found three marks on the go-fast's engine cowling that indicated something had struck it and glanced off. Only one round had pierced the cowling, and even that was stopped before doing any further damage. The fifth round was discovered inside the barrel of the shotgun.

"We determined there was enough water going into the gun that the ammo was getting soaked and causing the powder in the shotgun rounds to not fully expend," Laufenberg concluded. "Had the rounds operated as designed, that boat would have been stopped."

In this case, the Coast Guard crew had already been operating in the red due to

---

8 Double-aught buckshot is used for hunting deer and large game. A shotgun can send nine large-caliber pellets at over 1,300 feet per second; they are powerful enough to go through car doors and metal panels.

the dangerous nature of the go-fast chase; departing from the established warning shot/ disabling fire protocol made the situation much worse. Everyone involved from the boat crew to the regional command in Key West had succumbed to tunnel vision. They were so focused on stopping the go-fast that their judgment became clouded. We all learned a valuable lesson: never let the heat of the moment affect your ability as a leader to think carefully about your decisions. This incident is a clear example of what not to do when risk-management techniques show an operation is in the red zone.

Another incident later in the year was an example of the right way to operate. A few weeks prior to the U.S. presidential election during the fall of 2004, Station Marathon spotted a slow-moving motorized chug with approximately seven Cubans aboard, a few miles offshore Grassy Key. We dispatched the ready boat from Station Marathon with BM2 Joel Laufenberg and BM3 Joshua Soto and two others. Because the chug was so near the shore, I grabbed a crew also and manned another boat to the scene. Under the "wet foot, dry foot" policy, reaching shore was like reaching home base. The migrants would become combative close to shore and often carried crude weapons such as machetes and knives, containers of gasoline for Molotov cocktails, and even loaded handguns.

BM2 Laufenberg made various attempts to stop the chug using nonlethal means, including an RGES—a device that propels a net to entangle a propeller. The migrants had greased up their bodies with Vaseline to make pulling them off the chug difficult, and they were standing up and angrily shouting. One of the migrants warned the Coast Guard to stand off or he would shoot.

"He was holding his arm outstretched toward the Coast Guard vessel with his hands inside a bag," Laufenberg later recounted. "One of my crew members translated what the migrant was saying. He stated that he had a gun in the bag, and he was going to shoot us. Everyone on board took cover and drew their weapons. As the coxswain, I was unable to seek cover behind anything or draw my weapon."

BM2 Laufenberg pulled the Coast Guard vessel back to evaluate the situation. They were already operating in the red due to the combative migrants. Normally the Coast Guard vessel would move in hard and fast, and the crew would physically pull the migrants off the boat and put them in cuffs if they resisted.

At the time, the Coast Guard's policy on the use of force allowed crew members to use deadly force if three elements were present: ability, opportunity, and jeopardy/intent. In this case, the migrant clearly had the ability to use deadly force (assuming he had a gun in his bagged hand), he had opportunity because he was pointing it at the Coast Guard, and his shouts and aggressive demeanor clearly communicated his intention to shoot. BM2 Laufenberg and his crew would have been operating completely within policy if they had shot the migrant.

BM3 Soto, located near the front of the response boat, elected not to shoot—he didn't believe the man had a weapon. The crew crouched low below the Coast Guard vessel's sides, providing a smaller target and some ballistic protection from handguns. Knowing that a second Coast Guard boat was approaching, the crew held fire and moved back from the chug.

As I arrived on scene with the second response boat, we did a quick risk-management evaluation. I agreed with BM3 Soto's assessment that the migrant was bluffing. We were now less than 300 yards from shore, and if we were going to take action we needed to do it immediately. We decided on a high-speed pincer movement with both Coast Guard vessels coming at the chug from separate sides, slamming into it simultaneously. We had designated shooters on both vessels who were watching all the migrants closely for any indication that one of them had a gun and would shoot it.

Once alongside the boat, we first pepper-sprayed the migrant holding the bag, which did, indeed, turn out to be empty. Other migrants waved sticks with nails at us, and we pepper-sprayed them as well. We pulled all the migrants onto the Coast Guard vessels and cuffed them, then searched the chug. We found no firearms on board or on any of the migrants. If BM3 Soto had fired when the migrant shouted that he had a gun, his actions would have been within policy, even though no gun was ultimately found. Still, the migrant would be dead, and the political ramifications could have been enormous. Shooting an unarmed Cuban migrant a few weeks before the presidential election could have shifted the support of the Cuban-American population in South Florida from President George W. Bush to his opponent. And if Bush had lost Florida, he would have lost the election, thereby changing American history. Beyond the life-and-death implications of this case, this example showcases how important a single decision can be.

Despite clearly operating in the red—and in a chaotic situation—the boat crew approached each decision carefully and stayed within policy. No matter how hectic operations on the scene became, the crew did not succumb to tunnel vision or attempt to force the situation beyond established protocol. When events started getting out of hand, the boat crew thought through their next course of action. They did not shy away from taking action, but they did deliberate. This mission's outcome was much different from the earlier one with the same crew. The lesson was clear: do not hesitate to operate in the red zone if the importance of the mission dictates, but pay close attention to staying within policy and standards. ∎

Lesson Five: Communicate Effectively

Nothing lowers the level of conversation more than raising the voice.
~ Stanley Horowitz

Without effective communication, every other principle of good leadership is useless. Verbal skills, clear orders, and written instructions are all essential to connecting with your team and conveying your ideas, but they're just the beginning—demeanor, deportment, and how you live your life are fundamental ways to communicate direction, intent, and management. Occasionally, good communication calls for thinking outside the box and using creative means to deliver your message to the crew. Communication should be tailored not only to fit specific situations, but also to fit specific personalities. Communication that works for one individual might be entirely ineffective on another.

Treat your crew and command cadre the way you want them to treat their subordinates—behavior is an invaluable form of communication. Your team observes everything you do. If I ever witnessed a crew member's inappropriate verbal tone or action, I would immediately have a private one-on-one counseling session with them.

"The day you see or hear me act like that toward you is the day you can speak and act that way to your fellow crew members," I would say.

There is no exact way to convey this, and every crew and situation is different. But if you use this type of approach as you form your own leadership style, you will definitely have a positive impact on your team. A strong team equals mission success almost every time.

One incident in the summer of 1998 illustrates the importance of communication through behavior. My crew and I had traveled to Louisiana to pick up the newly constructed Coast Guard Cutter *Hammerhead* (WPB 87302) from Bollinger Shipyards. I knew the crew well—most had served with me on Coast Guard Cutter *Point Turner* (WPB 82365) until her decommissioning. We were on a new cutter, but apart from a few new additions, we had been working together as a team for over a year. Two of our new additions were BM1 Andrea Currie, executive petty officer, and QM2[9] Elena Soini, operations officer.

We spent the summer and quite a bit of the fall in Louisiana learning how to operate the new vessel before we took it up north to its new home in Woods Hole, Massachusetts, on Cape Cod. The *Hammerhead* was the second cutter of a new class of 87-foot, high-tech patrol vessels that boasted the very best in advanced communications and electronics on the bridge, along with a smaller boat that could be launched directly from the stern of the cutter while it was underway.

By December, we were conducting some of our first patrols off the coast of New England. Our operational area included many large areas offshore that were closed to fishing

---

[9] Quartermaster 2nd class

and scalloping, so one of our missions was to conduct boardings and inspections to enforce federal fishery regulations. One evening, we came upon the fishing vessel Liberty, which was dragging for scallops near one of the closed areas on Georges Bank, about 80 miles offshore of Nantucket. We decided to board the vessel. The seas were running between three and six feet, visibility was good, and the wind was blowing about 15 to 20 knots.

The Hammerhead's small cutter boat was a new vessel designed and manufactured specifically for use with the innovative launching feature. It was 17 feet long with a jet drive, so you could power the vessel right up into the stern notch of the 87-footer. Before the creation of the new 87s, one of the most dangerous routine evolutions on small cutters was launching the small boat, a process that had always required cranes to lift the small boats off the main deck of the ship and into the water. In rough seas, it was easy to lose control of the small boat as you were trying to maneuver it. On various vessels and several occasions throughout my career, lines snapped or booms failed while launching a small boat, presenting serious danger to the crew. The stern launch system was developed to avoid that hazard, as well as to save valuable time during missions.

We launched the Hammerhead's small boat with a crew of four, including BM1 Currie, MK2 Kenny Yarbrough, SN[10] Jovan Bielski, and FS2 Brian Cassetta. The night was pitch black with only the bright deck lights of the fishing vessel in sight. The crew members of the small boat placed themselves alongside the Liberty, and three of them moved forward to climb up and board the vessel on its port side. As they moved to the bow, a large wave caused the Liberty to suddenly roll starboard. The small boat abruptly flipped over, displacing its crew into the frigid Atlantic Ocean. I was on the bridge of the Hammerhead with QM2 Soini when it happened.

"I watched the launch and transit and was using binoculars to watch their approach," she recalled. "I could see nav lights occasionally and the retro on the crew helmets as they approached the F/V,[11] and then I saw nothing. I said, 'Master Chief, where's our small boat? I don't see it!' It was the only time I ever heard him swear."

The crew of the Liberty shouted over the radio that our small boat had overturned, while a second radio began transmitting the same urgent message from the downed crew. I attempted to activate the cutter's suddenly faulty spotlight to no avail. QM2 Soini immediately alerted the rest of the crew who were below deck on the Hammerhead, then began flipping breakers in the dark to get the spotlight working.

As my crew quickly gathered on the main deck, I told the Liberty to move out of the way so I could get to the overturned small boat. I barked out orders to my crew, instructing them to get heaving lines and flotation devices ready. Responding to the elevated volume and tension in my voice, they moved faster and more efficiently than they probably ever had before. My crew had never heard me raise my voice like that, and the change instantly conveyed the severity of the situation.

What had seemed to be a normal sea state only moments before now appeared much more violent as the wind gusted higher. I maneuvered the 87-foot cutter toward the overturned boat and immediately saw that I only had two choices: I could drift down on

---

[10] Seaman
[11] Fishing vessel

the small boat, letting the winds push me (usually the best course of action when people are in the water), or I could come into the wind and pick them up from the windward side. I decided to come into the wind. The overturned vessel's profile was so low in the water, I was afraid I'd drift completely over the awash vessel if I let the wind push me. Coming against the wind offered me more control. We had just a few minutes to pick up our crewmates before hypothermia set in; they were only wearing full-body flotation suits that weren't designed to fully preserve body heat.[12]

I pushed into the wind and attempted to get near enough to the small boat so the crew could throw heaving lines over. As I got closer, I could only see three heads around the bow of the small boat. BM1 Currie's right leg strap had been caught up in the *Liberty*'s gear, and she was pulled down several feet before it released. She came up to the surface under the overturned boat and was now trapped in a pocket of air, but the buoyancy of the flotation suit prevented her from getting out. The wind pushed the *Hammerhead* away, and I had to reset the cutter to come in at a sharper angle. While I was doing that, BM1 Currie was sweeping her legs under the overturned boat to alert her crewmates. She caught MK2 Yarbrough's leg, grabbed it with her hand, and gave four short tugs. Yarbrough and the other two downed crewmen worked together to overpower her flotation suit, shove her down, and pull her out from under the capsized boat.

By this time, the *Hammerhead* was coming in at an extremely sharp angle. SN Bob Parent was able to connect with the life ring—no easy feat since he was throwing into the wind—and we threw heaving lines to the four crew members and pulled them to the side of the cutter. By the time they were on the deck, almost 15 minutes had passed since they had gone in the water, and they were already hypothermic. While our deck crew tended to them, I turned the cutter to look for the small boat. We continued searching for the rest of the night, but it wasn't until daybreak that we finally found it.

A few years later, Andrea Currie sent me a card containing a photograph of the *Liberty* and a quote from oceanographer R. M. Snyder:

> There are things about the sea which man can never know and never change. Those who describe the sea as "angry" or "gentle" or "ferocious" do not know the sea. The sea just doesn't know you're there—you take it as you find it, or it takes you.
>
> Remember always our mortality,
> BMC Andrea Currie

I've often thought about that night off the New England coast. In fact, I had the photo and card framed and have kept it on the wall of my various offices ever since. To me it is a reminder of how quickly the most routine situation can become life threatening, and how every single day that we're allowed to breathe the air is a blessing, not a guarantee. Everyone involved in this incident made it back to shore safely, thanks to clear, concise, effective communication. From QM2 Soini's notification to the crew that an emergency existed, to my abrupt orders shouted out of the pilot house to the crew on the main deck, to BM1 Currie remaining calm while trapped under the small boat and having the

---

[12] Dry suits later became mandatory in the Coast Guard

presence of mind to grab her crewman's leg, to MK2 Yarbrough keeping the crew together with the small boat and not trying to swim to the *Hammerhead*, to Yarbrough's ability to communicate to SN Bielski and FS2 Cassetta what they needed to do to pull Currie out from under the cutter boat despite the treacherous, freezing conditions—all these actions constituted efficient, effective communication.

> *In many ways, effective communication begins with mutual respect,*
> *communication that inspires, encourages others to do their best.*
> ~ Zig Ziglar

Communication styles can also be situational. The way I communicated with my crews on day-to-day operations was entirely different from the way I communicated with them during an emergency situation. When a leader relies on berating or screaming at his crew during every routine operation, it not only chips away at morale, it also quickly loses its effectiveness. When I worked for the chief who screamed at us all the time, it got to the point where I started to tune him out, because his screaming had become the status quo. When I reached a leadership position myself, I made a point to communicate not just by speaking but by listening to my crew as well. We maintained a dialogue, and I took their concerns and feedback into account when I made decisions. My crew knew they could talk to me, and if I ever did find the need to raise my voice, it was because the occasion called for it and demanded their immediate action.

> *The great enemy of communication is the illusion of it.*
> ~ William H. Whyte

During my last four years on active duty, I was honored to have Adm. Thad Allen (Coast Guard commandant 2006–2010) choose me to serve as master chief petty officer of the Coast Guard (MCPOCG). I had served with Admiral Allen on previous assignments and knew what a dynamic, inspiring leader he was. His service as the principal federal official during the response to both Hurricane Katrina and the Deepwater Horizon oil spill cast him onto the worldwide stage as a preeminent expert of emergency response and risk mitigation. The national media dubbed him "the master of disaster," but those of us fortunate enough to work with him closely knew him as a driven leader who passionately looked for ways to improve the world around him. If I had to pick one attribute of his that constantly impressed me, it would be his ability to communicate and develop a strong and sincere connection with everyone he met. He could connect equally with a head of state or the lowliest seaman apprentice in the service.

As the senior enlisted advisor to the commandant of the Coast Guard, my duties as MCPOCG included representing the commandant to the Coast Guard workforce, representing the workforce to the commandant and Coast Guard leadership, acting as "troubleshooter" for the issues within the workforce, and advising leadership on policy. I always considered myself, first and foremost, a "field reality check." Admiral Allen inspired everyone around him to work issues to make the service better, and I took full advantage of that empowerment to take the initiative on many projects.

One example of Admiral Allen's leadership style occurred early in my tour as

MCPOCG, during the fall of 2006. The Coast Guard Cutter *Healy* (WAGB 20), operating in the Arctic, experienced two deaths on board during a cold-water diving evolution. Lt. Jessica Hill, 31, of St. Augustine, Florida, and BM2 Stephen Duque, 26, of Miami, both drowned during what should have been a relatively safe training exercise. Along with a complete analysis of the Coast Guard Diving Program, Admiral Allen asked senior leadership to look broadly at Coast Guard activities and policies for anything we might be overlooking that could result in preventable accidents.

After a lot of thought, I focused on the Coast Guard's "egress" training program. Escaping from inside an overturned boat or ship is called "egressing" the ship. The maneuver is extremely difficult. Imagine the room you're sitting in suddenly turns upside down, and you're submerged in cold water and total darkness with debris floating in the way as you fight to find a way out. As Andrea Currie learned on the *Hammerhead*, getting out from under an overturned boat can be difficult even if the vessel is small. The only egress training most Coast Guard personnel received was some variation of taking crew members to a random compartment, blindfolding them, spinning them around, and having them attempt to find a way out. This was wholly inadequate. I was reading through Coast Guard accident reports when I noted that a member of an Alaska-based unit had nearly drowned when the 25-foot response boat had overturned and he had difficulties egressing. The Coast Guard's insufficient training in this area was a problem that had been overlooked, and I believed it would lead to another tragedy.

I found a partial solution in the Coast Guard's project to build a modern aquatic training facility for our aviators and rescue swimmers at the Aviation Technical Training Center in Elizabeth City, North Carolina. Part of the project called for a "dunker" unit for aviator egress training; Coast Guard aviators were trained to escape from a cockpit mockup module submerged and overturned in a pool. Research showed that aviators who had been through this type of training had a much higher chance of egressing their aircraft and surviving an aviation mishap.

I thought, *If we're training aviators to egress, why not boat crews?* The Coast Guard's boat-force training was conducted in Yorktown, Virginia, about 90 minutes away from the new aquatic training facility. Why couldn't we transport boat-force students to the Elizabeth City site and have them go through egress training with a module configured to mimic the conditions in an overturned vessel?

I approached leadership of the aviation and boat forces programs with my proposal and was confident I had their support. Plans for the vessel module and transportation logistics got underway, and I thought we had the resources to make it happen. Sometime that fall, however, I learned that the budget for the pool was coming up $3.5 million short, and project leaders had proposed dropping the egress trainer from the project. I was upset and said so to the commandant's executive assistant, Capt. Tom Ostebo. As an aviator, he knew the value of the dunker training and suggested I bring up the issue at the next weekly flag meeting, when Admiral Allen met with all the local flag officers.

At the next meeting, I came prepared. My seat was near the head of the table in the Coast Guard situation room, so as the discussion moved around the room, my turn to bring up issues came quickly. I launched into an impassioned argument about the issue, how I had worked a deal between aviation and boat forces on the pool, and how if we didn't add $3.5 million to the aquatic training facility budget, Coast Guard people were going to die for lack of egress training. As I looked around the room, I saw blank looks on most of the flag officers' faces. It struck me that the one thing I had neglected—in all my preparation and passion—was to discuss the issue with the people who would have to find the money to pay for the dunker.

The vice admiral in charge of the Coast Guard's overall budget and resources looked upset. I had not even talked to Admiral Allen about the issue. Fortunately, despite my slip-up with communications, he did not abandon me. He calmly turned to the chief of Coast Guard engineering and asked what was going on with the project. The chief engineer admitted that the Coast Guard was $3.5 million short and that the egress trainer was in danger of being cut. Admiral Allen asked him to look into it.

I spent the rest of the day doing what I should have done before the meeting: socializing the issue with key leaders. That evening at 9:30, I received a four-word email from Admiral Allen: "We need to talk." I called him right away and asked, "Is this about this morning?" He responded, "Yes. Wrong time, wrong issue for that forum."

But then he took the blame for not spending enough time together; he said it was hard for us to remain in sync when we were always on opposite sides of the country. He explained that the previous evening he had led a meeting regarding an Office of Management and Budget proposed budget reduction of almost a billion dollars. The leadership was still trying to figure out how to deal with that the next morning when I walked in and basically demanded that if we didn't come up with another $3.5 million for the dunker, people were going to die. I could see, in retrospect, why my passionate speech hadn't gone over so well.

"You might win this one, Master Chief," the admiral told me, "but you may not win many more." We agreed to stay closer in sync.

Over the next several months, with Admiral Allen's assistance I raised awareness of the need for the dunker. We took the chairman of the Coast Guard's Congressional subcommittee[13] to tour the Elizabeth City site, so he could see for himself how inadequate existing facilities were. In the end, we got the $3.5 million for the dunker added back in, and the completed project featured a 50x25-meter rescue-swimmer pool with wave generators and an underwater modular-egress training pool.

Two leadership lessons stand out from this experience. The first is that communication can make the difference between success and failure. Had I made sure that everyone involved with the budget was fully aware of the issue, things would have gone much more smoothly. The second lesson is that Admiral Allen did not abandon me when I hit some bumps in the road. He had given me free rein to act, and he continued to support me and work with me to make the rudder adjustments we needed to keep the project moving forward with a successful conclusion. ■

---

[13] The Subcommittee on Coast Guard and Maritime Transportation is a subcommittee within the U.S. House Transportation and Infrastructure Committee. The Subcommittee has jurisdiction over maritime safety, security, law enforcement, and defense.

Lesson Six: Tenacity

Never confuse a single defeat with a final defeat.
~ F. Scott Fitzgerald

Being a good leader often means being able to recognize areas in your organization that could benefit from improvement, and then implementing the necessary changes. Identifying opportunities to enact positive change, however, is merely the first step toward actually making a difference in an organization—especially one mired in bureaucracy. The road from opportunity to innovation is often a long and difficult one, and tenacity is the vehicle that will get you from one end to the other. The best ideas are often abandoned long before reaching fruition, simply because someone threw in the towel when they met resistance, instead of pushing just a few more inches and experiencing a breakthrough.

At times in my career, it seemed as though every effort to make a positive change was met with a roadblock. One of those times was in 2003, when I was the CMC for Coast Guard Headquarters and saw the possibility to improve our law enforcement and security mission capability while also creating new career opportunities for Coast Guard members.

Since my early years in the Coast Guard, I believed the service should do a better job of supporting and sustaining expertise in the area of maritime law enforcement and security. I had found that training programs devoted to this mission set needed to be improved and professionalized. Formal training was only a few weeks long, and those who were ultimately certified to conduct the law enforcement mission were not even required to attend the training. Coast Guard members only needed to demonstrate performance standards to superiors through an on-the-job training program. I strongly felt this type of system could not create and maintain the expertise needed to professionally conduct the Coast Guard's law enforcement mission—one that seemed to grow more complex every year.

I believed the answer was to create a specific law enforcement occupational specialty or rating. The Coast Guard manages the bulk of an individual member's technical competencies through various ratings that allows them to streamline each member's training according to their specific needs. Because there was no law enforcement rating, members across several different ratings were called upon to complete law enforcement training and develop proficiency in an increasingly complex law enforcement mission set. After a few years, that training was wasted as the members moved in and out of units that actually conducted the law enforcement/security mission, or were promoted up the ranks and their duties shifted. In addition, I saw a real need to develop and foster professional maritime law enforcement career paths. Through a rating, members could retain expertise and build on it as they moved from place to place and up through the ranks. Law enforcement could be the main focus of an entire career, rather than a part-time additional duty.

Unfortunately, not everyone shared my concern. In fact, most senior decision makers did not see this as an issue at all. At the time, if you asked a senior Coast Guard member what they thought should be done to improve the Coast Guard's effectiveness in the law enforcement mission field, you would likely receive anecdotal stories about that person's own experience in law enforcement, like that time in 1977 they spent hours in the trawler's hold counting fish for the fisheries law enforcement mission. Most senior leaders simply didn't see this as the big-picture issue that I did, but I strongly believed it was much too important to ignore.

I joined forces with two other MCPOs, Jeff Smith and George Ingraham, to write a white paper recommending the creation of a law enforcement rating. The paper addressed the changes in the nation's security environment and detailed how those changes had affected the scope and complexity of the Coast Guard's law enforcement mission. The white paper and a follow-up study helped prompt the creation of the Advanced Law Enforcement Competency (ALEC) program. ALEC was designed to keep personnel trained in high-end tactics and procedures within the law enforcement mission area.

Though the program didn't create a new rating and therefore didn't address the question of how members could be fairly advanced, it seemed like its creation was at least a partial solution to the problem. The feeling of success was short lived. The program was never allowed to get off the ground, due in part to the obstructionism caused by senior-level people who had been tasked with its implementation. Even if ALEC had been fully launched as it had been intended, ensuring higher-level training and follow-on assignments, it still would have fallen short of resolving the full scope of the problem.

In 2006, I visited many of the Coast Guard's relatively new Deployable Specialized Force units, including the high-end Maritime Security Response Team (MSRT) in Chesapeake, Virginia. The unit's main mission was focused on counter-terrorism, but the building blocks needed to grow their expertise were related to security and law enforcement. During a unit all-hands call, I found that many of the members lacked the training and expertise in the unit's mission to become certified as team members. The Coast Guard's personnel system of constantly transferring members and moving them in and out of the mission sets had caused the unit to be so degraded operationally that it would not have been able to respond effectively to a law enforcement situation.

This prompted me to again push for a rating dedicated to these types of skillsets. Soon after, the Law Enforcement, Tactical, and Security Group Occupation (LETS-GO) study team, led by Rear Adm. Tim Riker, was stood up to recommend a way forward. The team worked extremely hard on the initiative for a year and a half, ultimately identifying approximately 1,500 billets or job positions to be converted to the new rating.

When the team briefed the vice admiral who was the Coast Guard chief of staff at the time, he came out strongly against the changes. He did not believe a rating that could possibly convert up to 1,500 positions or billets from their original ratings was warranted. His negative reaction could have killed the entire initiative, but I was so confident that a new rating was the only solution that I continued to press forward. In hindsight, I learned a valuable lesson that day. I had spent all my time and energy working at lower levels to help push the new rating forward, when I should have been socializing the proposed rating with the resource decision makers and the chief of staff.

That mistake set us back about a year, and we had to jumpstart our initiative almost from the ground up. Finally, our efforts paid off. In 2008, Admiral Allen sent an all-hands email announcing the creation of a law enforcement/security-focused rating: the Maritime Enforcement (ME) rating. I wrote in my accompanying blog post, "This is a major step forward for both the Coast Guard's ability to sustain expertise in this increasingly complex mission area, and for individual Coast Guard members who have sought to focus on the law enforcement/security mission area for a career but were not able to because there was no dedicated rating."

If I had given up every time I was told "no," I would be looking back on a much different career. When I decided a battle was important enough to fight, I dug in and pressed on despite all the pushback I got. As Admiral Allen used to say, "You have to decide on which hill you're willing to die."

> *Through perseverance, many people win success*
> *out of what seemed destined to be certain failure.*
> ~ Benjamin Disraeli

In May 2009, Admiral Allen and I spoke at a memorial service for Coast Guard hero Bernie Webber in Wellfleet, Massachusetts. Almost 60 years before, on February 18, 1952, Boatswain's Mate 1st Class Webber had been the coxswain of a 36-foot motor lifeboat that responded to reports that an oil tanker, the SS *Pendleton*, had broken in half during a fierce nor'easter. The bow of the vessel had sunk, taking the lives of the captain and seven crew members and leaving the stern of the tanker and the remaining 33 crew members to face the same fate. Braving 70-knot winds, driving snow, and monstrous 60-foot seas, Webber and a small crew of three other Coast Guardsmen navigated beyond the Chatham bar, and against all odds, found what was left of the *Pendleton*. He masterfully maneuvered his small vessel alongside the tanker over 30 times to avoid being smashed against the tanker's hull, and rescued 32 men one by one, losing only one to the raging seas. The rescue is still considered the greatest surface rescue in Coast Guard history.[14]

It was during his memorial service that I started thinking about Coast Guard heroes like Bernie Webber, and how they deserved to be honored, their actions remembered and used to inspire the generations of Coast Guardsmen who came after them. At the time, the Coast Guard was preparing to launch a new class of vessels called fast-response cutters, and I realized we were missing a great opportunity to name these new cutters after enlisted heroes. In fact, after the service I told Andrew Fitzgerald, the last surviving crew member from the *Pendleton* rescue, that I was going to try to get a Coast Guard cutter named after Bernie Webber.

During the flight home, I discussed with Admiral Allen my belief that the Coast Guard had not focused enough on history, heritage, and service ethos. I pitched the idea of naming the new class of cutters after enlisted heroes. Up to that point, most cutters named for people were named after senior government or high-ranking Coast Guard officers. I explained that naming the vessels after heroes who started out in the enlisted ranks would

---

[14] In 2016, Disney produced the movie "The Finest Hours," a portrayal of the dangerous rescue, directed by Craig Gillespie. The script for the movie largely stuck to its source material, Michael J. Tougias and Casey Sherman's book, *The Finest Hours: The True Story of the U.S. Coast Guard's Most Daring Sea Rescue* (Scribner: New York City, 2010).

ensure that lesser known, but very important, Coast Guard men and women would be remembered and honored. Admiral Allen was strongly in favor of the idea, and encouraged me to present it to Coast Guard Headquarters' naming board.

I soon found out there was a complicating factor: the cutter names had already been decided and approved. They were to be named for various nomenclatures associated with the idea of "keeping the watch." The naming board had already approved the name "Sentinel" for the first of the class, and subsequent vessels had similar monikers. Despite the commandant's support, when I suggested the idea to name the new cutters after enlisted heroes, the naming board voted against the change. Additionally, the Coast Guard's acquisition directorate was also opposed to the change, citing the potential difficulty of revising the cutter names after Congressional and other notifications had already been made and approved using the Sentinel names.

Despite the pushback, I continued to press for the name change and asked the various programs to reconsider. Getting the new class of cutters named for our service's heroes was important. It was a way to honor the selfless men and women who chose to put others' lives before their own, to go above and beyond their duty as Coast Guardsmen. It was important to keep the stories of their sacrifices alive, to inspire the men and women who are now serving, to pave the way for the next generation of heroes to follow in their footsteps. It was certainly too important to let the slight inconvenience of a little extra paperwork get in the way.

The nature of my work as MCPOCG often took me away from headquarters, making it impossible for me to attend every naming board meeting, so the deputy commandant for operations CMC agreed to advocate for me when I was unavailable. Unfortunately, despite our dogged efforts, the requests to reconsider went on for the better part of a year, with no change. In February 2010, just three months before my retirement ceremony, I realized that the various decision makers were simply dragging their feet until I was out of their way and the new commandant's leadership was installed. I was running out of time and would have to find another way to get the change pushed through.

I decided to use the "nuclear" option: I spoke to Vice Commandant Vice Adm. David Pekoske and asked him to intervene directly. Vice Admiral Pekoske personally ordered the entire fleet of 58 fast-response cutters to be named after enlisted heroes, ending all stonewalling and pushing the initiative over the finish line.

When I am long gone from the Coast Guard's memory, new members will report aboard the USCGC *Bernard C. Webber* or the USCGC *Charles David Jr*, or a multitude of others, and they will know the value these great heroes brought to the nation and the Coast Guard. I could have easily given up on my efforts at the first roadblock. I had done my due diligence by presenting the idea to the naming board and letting it go to a vote. I chose to pursue the issue because I knew how important it was to honor the legacy of these Coast Guard heroes. It didn't matter how many times we were told "no"—we succeeded with perseverance and tenacity. ∎

Lesson Seven: Transitioning

Life is pleasant. Death is peaceful. It's the transition that's troublesome.
~ Isaac Asimov

Retiring from the Coast Guard was very difficult for me. I thought I had prepared for it, but when it actually happened, it was quite a shock. One moment you're involved in everything the service is doing, and the next moment you're totally out. It was a very hard pill to swallow.

The first step out of my comfort zone was determining what I was going to do in the civilian world. I accepted a position at a professional services and information technology company as a senior program manager. My colleagues were very good to me and I was surrounded by retired and former military, but the transition was still challenging. Much of the work had to do with proposals and business development—neither of which was familiar to me.

After nine months, I accepted a position at Bollinger Shipyards in southeast Louisiana, first as program manager for the fast-response cutter (FRC), a brand-new set of ships, and then as vice president of government relations. I loved their vessels and had a lot of experience as a Bollinger customer—they built the *Farallon* and *Hammerhead*—so I jumped at the chance. I had a steep learning curve, though, as I had little engineering, contracting, or acquisition experience, all of which were important in the new position. With a lot of assistance from my co-workers at Bollinger and some additional contracting training, I was able to contribute to the major success of the FRC program.

Having been through this challenging experience, I can offer a few helpful tips to those who are or will be transitioning from the military to civilian life:

1. Get your degree
I can't overstate the importance of education. Shortly after I took my first civilian job, I found myself in charge of looking through résumés to fill a position. A lot of great people were in that pile, but I wasn't even able to consider those who didn't have a college degree. Many well-regarded, accredited universities are willing to award extensive credit for military training and experience. I am currently on the military team at Purdue University Global, a great university focused on helping service members achieve educational success. Getting a college degree should be at the top of your priority list—choose the school that will best fit your needs.

2. Get your PMP certification
If you can get your Project Management Professional certification while you're still on active duty, you're more likely to get a job with a government contractor.

3. Keep your security clearance current
Don't ever let it expire—you never know when it will come in handy.

4. Network, network, network
Your military skills and experience will be in high demand, but you have to let people know you're looking. Keep an open mind about every job.

5. Appreciate the advantages

Nothing will fully replace the adventure you had in the military. Maybe as time goes on it will get easier, but I will always miss the Coast Guard. Still, there's something to be said for living wherever you want and going home at night on a regular basis. Look for the advantages and positive aspects of working in the civilian sector, instead of focusing on the negative differences.

6. Fully prepare for the transition

The Coast Guard's famous motto is *Semper Paratus,* which means "always ready." If you think through every aspect of post-retirement life, you'll be ready when the time comes. Some time ago, I counseled a retired E-9 who had never completed his college degree and had let his security clearance lapse. To say that he had a difficult time in transitioning is a major understatement. Don't let it happen to you—anticipate what you'll need to do to be successful in your civilian life.

*   *   *   *   *

On May 21, 2010, my 32-year Coast Guard career came to a close in the same place it had started, at Recruit Training Center, Cape May, New Jersey. Taking a very short break from the Deepwater Horizon oil spill cleanup, Admiral Allen officiated. Congressman Frank LoBiondo also honored me by taking part in the ceremony, as did Pastor Conrad De La Torres, who came all the way from Homestead, Florida, and mentors such as former BMCM[15] Rocky Bucci and former MCPOCG Vince Patton.

I became vice president of government relations at Bollinger Shipyards soon after retiring from the Coast Guard, and I'm also a consultant for several other organizations. I am a proud member of the military advisory board for Veterans United, the largest (and the best, in my opinion) provider of VA home loans in the United States. I currently serve as president of the Association for Rescue at Sea, a nonprofit dedicated to supporting volunteer search and rescue efforts worldwide. Additionally, for the last six years I have been privileged to serve on the advisory board of First Command Financial Services, an organization dedicated to helping military members attain their financial goals. Previously, I served as co-chair of the Coast Guard's national retiree program and on the boards of the Coast Guard Enlisted Memorial Foundation, the Armed Forces Retirement Home, and Broward Navy Days.

My work on these various boards inspired me to form, along with five other former senior enlisted leaders, the consulting company Summit Six, LLC. Having reached the apex of our military careers, we are keen to use our experience and leadership skills to continue making a difference for our nation and our military.

I'll let others decide whether I was a successful leader or not. For my part, I tried to set an example at every command I had. I tried to be respectful and involved with each member of my crew, and I did my best to coach and mentor those crew members in their careers and military life. I provided an open-door atmosphere so they were free to interact with me, understand me, and, in a short time, respect me. There was never any doubt that

---

15 Boatswain's mate master chief petty officer

they were my priority. I had their backs, and to this day, I consider them family. I'm proud to say their lives and service in the Coast Guard are my legacy. ■

TOP LEFT: In exposure suit on boat cabin top, Station New Haven, Connecticut, 1992.
TOP RIGHT: Briefing in Baghdad, Iraq, 2007.
ABOVE: On dock in a storm, Station New Haven, Connecticut, 1993.
LEFT: Talking with Gen. David Petraeus, HQ-MNF, Iraq.

TOP: With Adm. Thad Allen during an all-hands meeting in Kodiak, Alaska, 2007.
ABOVE: At the White House with Michelle Obama during a "Women in the Service" event, 2009.
RIGHT: At the helm of a Coast Guard vessel, Station New Haven, Connecticut, 1993.

TOP: With Recruit Training Company commanders, Cape May, New Jersey, 2009.
ABOVE: With fellow service SEAs and President Barack Obama at the Commander in Chief's Ball, National Building Museum, Washington, D.C., Jan. 20, 2009. The ball honored America's service members, families, the fallen, and wounded warriors.
LEFT: With crew of Coast Guard Station Atlantic City, New Jersey, 2009.

Afterword

T his is an unusual afterword. Authors' spouses don't ordinarily write commentaries in books, but our perspective isn't ordinary. Military spouses do more than just hold the fort down while their husbands and wives are away, sometimes for months at a time; they uniquely serve as ground anchors for their service member spouses, providing those leaders the unhindered ability to give their full attention to serving our country. A solid, happy marriage—one of the many qualities these six authors have in common—can make the difference between a distracted, unhappy individual and a focused, positive leader. The six of us knew from the beginning of our spouses' military careers what sacrifices would be required, and we willingly took on the responsibility. After all, we love our country, too. It was a profound honor to support our spouses as they rose to become leaders in the U.S. Armed Forces.

For some of us, our adventures together began when we were just kids—several of us got married before we could even vote—so the demands of military life came gradually. One of our spouses proposed romantically by saying, "I deploy a lot—can you handle that?" Between the six couples, we've been married an astounding total of 201 years, making us fairly qualified to give a few tips on how to have a strong, happy marriage with a military spouse.

What did we do to support them? We loved them, prayed for them, encouraged them, understood their difficult schedules, saw moving every few years as an adventure, learned how to be independent, kept busy and happy, tried not to complain or be demanding, attended hundreds of receptions and conferences cheerfully, and, of course, did the lion's share of household management and child rearing. We also made sure they knew how proud we were of them.

That's not to say we were always model spouses. We're as human as anyone, and military life was often difficult and a burden. Even homecomings weren't always a honeymoon: when your spouse comes home after a yearlong deployment, he or she might forget that you've been the one in charge. You've been the one getting the brakes fixed, changing the swamp-cooler motor on the roof, attending the kids' school meetings, helping with homework and science fair projects, doing the taxes, mowing the lawn, cleaning up kid vomit and baby poop, taking the pets to the vet, putting up the Christmas tree, and doing all the shopping—and suddenly they think they're the boss, and they're disrupting your finely tuned schedule. It's tough to assimilate back into home life—for both spouses. We had to speak our minds honestly, but also be kind. The one at home and the one deployed had both had it hard, and we needed to communicate and understand each other. There was no room for resentment.

Even though we were not in the spotlight as our spouses were, we had important responsibilities in the military, behind the scenes. We were placed in positions to create positive change in the lives of our military families—a truly humbling experience. We mentored young wives and husbands and gave them a community that had their backs. We were front-row witnesses to the consequences of war, devastation, and broken families, and we did all we could to help those affected.

It's been an honor and a privilege to watch our spouses reach the highest enlisted levels in the military, knowing that we've been an integral part of their journey. Supporting our spouses and helping them become the leaders they are today has been our unique way of serving our country.

Karen Preston
Susan Barrett
Bobbi West
Paula Roy
Gary Hall
Janet Bowen

About Summit Six

The six authors of *Breaching the Summit*—each a retired senior enlisted advisor (SEA) to a member of the Joint Chiefs of Staff—founded Summit Six LLC, a consulting company, in 2017. Team members (pictured above from left to right) include Rick West, former Master Chief Petty Officer of the Navy; Mike Barrett, former Sergeant Major of the Marine Corps; James Roy, former Chief Master Sergeant of the Air Force; Ken Preston, former Sergeant Major of the Army; Denise Jelinski-Hall, former Senior Enlisted Advisor to the National Guard Bureau; and Skip Bowen, former Master Chief Petty Officer of the Coast Guard. Summit Six provides world-class transformational leadership coaching, organizational-change management, workforce development and alignment, and motivational speaking solutions. The group's mission is to focus on improving workforce leadership ability by using the skills, techniques, and experiences gained from military service, post-military corporate careers, hard work, commitment, integrity, and respect for others. ■

For Further Reading About the Authors

Kenneth O. Preston

Sean Simmons, "SMA Preston Gets Face to Face With 2nd BCT Soldiers," official website of Fort Riley and 1st Infantry Division, May 9, 2007, http://www.riley.army.mil/News/Article-Display/Article/470062/sma-preston-gets-face-to-face-with-2nd-bct-soldiers/.

Jacqueline M. Hames and C. Todd Lopez, "Longest-serving SMA Says Goodbye," Fort Campbell Courier, Feb. 10, 2011, http://fortcampbellcourier.com/news/article_c7087bb2-3559-11e0-bf51-001cc4c002e0.html.

Kerry Otjen, "Sergeant Major of the Army Reflects on Army's Growth," Fort Campbell Courier, Feb. 24, 2011, http://www.fortcampbellcourier.com/news/article_f8737814-405d-11e0-8626-001cc4c002e0.html.

C. Todd Lopez, "Preston Retires After Record SMA Stint," official website of the U.S. Army, Mar. 1, 2011, https://www.army.mil/article/52612/preston_retires_after_record_sma_stint.

Nicole Leighty, "Sergeant Major of the Army Kenneth Preston Visits Frostburg State," The Bottom Line, Apr. 15, 2017, http://thebottomlinenews.com/sergeant-major-of-the-army-kenneth-preston-visits-frostburg-state/.

Micheal P. Barrett

"DADT: Micheal Barrett, Sergeant Major of Marine Corps, on Gay Rights," Huffington Post, June 21, 2011, http://www.huffingtonpost.com/2011/06/21/michael-barrett-dadt-gay-rights_n_881109.html.

"Marines: Next Enlisted Leader 'Like Superman'," Military Times, Mar. 22, 2013, http://www.militarytimes.com/2013/03/22/marines-next-enlisted-leader-like-superman/.

Arthur P. Brill Jr., "SgtMaj Micheal P. Barrett, 17th Sergeant Major of the Marine Corps, At the Singleton Distinguished Lecture Series," Leatherneck, Sept. 2013, Vol. 96, Issue 9, https://www.mca-marines.org/leatherneck/2013/09/sgtmaj-micheal-p-barrett-17th-sergeant-major-marine-corps-singleton.

Walter Pincus, "Marine Corps Sgt. Maj. Micheal Barrett Takes Heavy Flak for Capitol Hill Testimony," The Washington Post, Apr. 14, 2014.

Sara W. Bock, "Outcall with the 17th Sergeant Major of the Marine Corps," Leatherneck, Feb. 2015, Vol. 98, Issue 2, https://www.mca-marines.org/leatherneck/2015/02/outcall-17th-sergeant-major-marine-corps.

Rick D. West

Bill Houlihan, "MCPON Stresses Brilliant on the Basics to NC Symposium," official website of the U.S. Navy, June 19, 2009, http://www.navy.mil/submit/display.asp?story_id=46356.

LaTunya Howard, "Going the Extra Mile to Welcome New Sailors," The Flagship, Dec. 3, 2009, https://www.militarynews.com/norfolk-navy-flagship/news/from_the_fleet/going-the-extra-mile-to-welcome-new-sailors/article_91bca8fa-f315-5206-81ea-d1c9363051a9.html.

Maria Yager, "CNO, MCPON Meet Sailors in Afghanistan," official website of the U.S. Central Command, Jan. 23, 2011, http://www.centcom.mil/MEDIA/NEWS-ARTICLES/News-Article-View/Article/884220/cno-mcpon-meet-sailors-in-afghanistan/.

Thomas L. Rosprim, "MCPON Testifies Before Congress on Hazing," The Flagship, Mar. 28, 2012, https://www.militarynews.com/norfolk-navy-flagship/oceana/news/mcpon-testifies-before-congress-on-hazin/article_0e0da6f9-1134-5be9-b5bd-98a3fe797228.html.

Rear Adm. James Foggo, "A Farewell to the MCPON," U.S. Naval Institute Blog, Sept. 28, 2012, https://blog.usni.org/posts/2012/09/28/a-farewell-to-the-mcpon.

James A. Roy

Peter Grier, "Chief Roy," Air Force Magazine, Dec.2009, http://www.airforcemag.com/MagazineArchive/Pages/2009/December%202009/1209chief.aspx.

Michael L. Howard and Rachel L. Griffith, "Air Force View: The Word from the Top," Army Space Journal, 2011 Spring/Summer edition, https://www.yumpu.com/en/document/view/26337949/spring-2011-space-and-missile-defense-command-us-army/26.

Ian Hoachlander, "Air Force's Top Enlisted Leader Visits JB Charleston," official website of Joint Base Charleston, Apr. 10, 2012, http://www.jbcharleston.jb.mil/News/Article-Display/Article/233847/air-forces-top-enlisted-leader-visits-jb-charleston/.

Mark Thompson, "Common Sense from an Enlisted Man," Time, July 20, 2012, http://nation.time.com/2012/07/20/common-sense-from-an-enlisted-man/.

Chris Powell, "Carrying the Message: CMSAF Provides Vital Link Between Enlisted Airmen, AF Leaders," Airman Magazine, Sept. 10, 2012, http://www.onemilitaryday.com/2012/09/10/carrying-the-message/.

Denise M. Jelinski-Hall

Ellen Krenke, "Guard Senior Enlisted Advisor Celebrates History of American Fighting Women," official website of the U.S. Air Force, Nov. 12, 2010, http://www.af.mil/News/Article-Display/Article/114994/guard-senior-enlisted-advisor-celebrates-history-of-american-fighting-women/.

Terry Lehrke, "Raised with Strong Faith and Foundation of Values, CMSgt Denise Jelinski-Hall Oversees Entire Enlisted Corps for National Guard," Morrison County Record, Feb. 23, 2012, https://www.hometownsource.com/morrison_county_record/news/local/raised-with-strong-faith-and-foundation-of-values-cmsgt-denise/article_45ac854d-75ca-53e3-b1c0-9aab60c24705.html.

Claudette Roulo, "Guard Enlisted Leader Stresses Support Available," American Forces Press Service (U.S. Dept. of Defense), Sept. 28, 2012, https://www.military1.com/army/article/284-guard-enlisted-leader-stresses-support-available/.

Jim Greenhill, "Jelinski-Hall: Men, Women United by Uniform, Single Standard," official website of the U.S. Army, May 29, 2013, https://www.army.mil/article/104389/jelinski_hall_men_ women_united_ by_uniform_single_standard.

Interview with Denise Jelinski-Hall, official website of the U.S. Army Educational Package promoting the film "Unsung Heroes: The Story of America's Female Patriots," 2016, http://www.unsungheroeseducation.com/Denise-Jelinske.html.

Charles W. "Skip" Bowen

Interview with Charles "Skip" Bowen, "All in a Day's Work for MCPOCG Bowen," Naval Affairs Magazine, Aug. 2006, Vol. 85, No. 8, https://www.fra.org/FRA/Web/FRA_Docs/FRAToday/2006/August2006/August%202006.pdf.

Bill Houlihan, "MCPON, MCPOCG Attend Senior Sailor Symposium in New Zealand," official website of the U.S. Navy, Nov. 11, 2007, http://www.navy.mil/submit/display.asp?story_id=33452.

Bill Houlihan, "MCPON, MCPOCG Open Global Maritime Senior Enlisted Symposium," official website of the U.S. Navy, Sept. 23, 2008, http://www.navy.mil/submit/display.asp?story_id=39932.

Interview with Charles "Skip" Bowen, "Former Coast Guardsman Semper Paratus in Private Sector," Seniormilitaryintransition.com, Aug. 13, 2012, http://seniormilitaryintransition.com/former-coast-guardsman-semper-paratus-in-private-sector/.

Kristen N. Bowen, "Oral History of MCPOCG Charles 'Skip' Bowen," official media website of the U.S. Dept. of Defense, Oct. 26, 2017, https://media.defense.gov/2017/Oct/26/2001833653/-1/-1/0/BOWENCHARLESMCPOCG_ORALHISTORY.PDF.

To book speaking engagements
with any of the authors,
please contact us at
SummitSixTalk@gmail.com